Contents

1 Chapter 1: Earth's Fiery Heart

Our planet, vibrant and dynamic, pulses with a hidden energy deep within its core. This internal heat engine, a remnant from Earth's fiery formation, drives the very processes that shape our world, from the majestic rise of mountain ranges to the devastating tremors of earthquakes. Understanding this internal structure is crucial to grasping the forces that create both volcanoes and earthquakes. Imagine the Earth as a layered sphere, not unlike an onion. At its heart lies the inner core, a solid iron-nickel alloy subjected to immense pressure and temperatures rivaling the surface of the sun. Surrounding this solid heart is the outer core, also composed of iron and nickel, but in a liquid state due to slightly lower pressure. This churning molten metal generates Earth's magnetic field, a protective shield against harmful solar radiation.

Moving outward, we encounter the mantle, a thick, rocky layer comprising the bulk of Earth's volume. The mantle itself is subdivided into regions based on varying physical properties. The lower mantle, bordering the outer core, is relatively rigid due to intense pressure. The upper mantle, closer to the surface, is more plastic and can flow slowly over geological timescales. This region, known as the asthenosphere, plays a critical role in plate tectonics. It is here that the immense heat from the core causes convection currents, driving the movement of the tectonic plates that make up Earth's outermost layer. These plates, rigid pieces of lithosphere (which includes the crust and the uppermost part of the mantle), are constantly in motion, albeit at a glacial pace.

Finally, we reach the Earth's crust, the outermost layer, a thin and brittle skin compared to the layers beneath. This is the layer we live on, the solid ground beneath our feet. The crust is further divided into two types: oceanic crust, denser and thinner, and continental crust, less dense and thicker. It's the interactions between these plates, driven by the heat from below, that create the geological phenomena we experience as volcanoes and earthquakes. The constant movement of these plates, interacting at their boundaries, leads to the fracturing and melting of rock, ultimately forming magma. Magma, molten rock found beneath the Earth's surface, is the lifeblood of volcanoes.

The formation of magma is a complex process primarily driven by three mechanisms: decompression melting, flux melting, and heat transfer melting. Decompression melting occurs when hot mantle rock rises towards the surface, experiencing a decrease in pressure. This pressure reduction lowers the melting point of the rock, causing it to melt without any addition of heat. This process is particularly prevalent at divergent plate boundaries, where plates move apart, allowing mantle material to ascend. Flux melting, on the other hand, involves the addition of volatiles, such as water and carbon dioxide, to the mantle. These volatiles lower the melting temperature of the surrounding rock, triggering melting. This mechanism is common at convergent plate boundaries, where one plate subducts, or dives beneath, another. The subducting plate carries water with it, which is then released into the overlying mantle, initiating flux melting.

Heat transfer melting occurs when hot magma from the mantle rises and intrudes into the cooler crustal rocks. The heat from the magma is transferred to the surrounding crust, increasing its temperature and causing it to melt. This mechanism often contributes to the formation of large magma chambers beneath volcanoes. The composition of magma varies depending on the source rock and the melting process involved. Basaltic magma, rich in iron and magnesium, is typically formed by decompression melting at divergent boundaries and hotspots. Andesitic magma, with an intermediate composition, is commonly associated with flux melting at convergent boundaries. Rhyolitic magma, rich in silica, is often formed by heat transfer melting in continental crust. These compositional variations influence the viscosity, or stickiness, of the magma, which in turn affects the nature of volcanic eruptions.

Understanding the origins and properties of magma is crucial for predicting volcanic behavior and mitigating volcanic hazards. The interplay of these factors—the Earth's internal structure, plate tectonics, and magma formation—sets the stage for the dramatic events we witness at the Earth's surface. These processes, driven by the immense energy within our planet, shape the landscape and remind us of the dynamic forces at play beneath our feet. From the slow creep of continents to the explosive eruption of a volcano, the Earth's fiery heart is a constant reminder of our planet's dynamic nature.

1.1 The Planet's Interior

Our Earth, seemingly solid beneath our feet, is a dynamic entity with a fiery heart. It's structured like a colossal onion, composed of distinct layers, each with unique properties that contribute to the planet's overall behavior. At the very center lies the inner core, a solid sphere of predominantly iron and nickel, enduring immense pressure and temperatures comparable to the surface of the sun. Surrounding the inner core is the outer core, also composed primarily of iron and nickel, but in a liquid state. This molten layer, swirling and churning due to the Earth's rotation and heat from the inner core, generates the planet's magnetic field, a protective shield against harmful solar radiation. This churning also plays a crucial role in the transfer of heat to the next layer, the mantle.

The mantle represents the largest volume of the Earth, a vast expanse of semi-molten rock, predominantly silicates rich in magnesium and iron. Its consistency is neither fully solid nor fully liquid, instead resembling a highly viscous, plastic-like material. Within the mantle, we encounter slow, creeping currents known as convection currents. Heated material from the core rises, cools near the surface, and sinks back down, driving the movement of the tectonic plates that make up the Earth's outermost layer. These convection currents are the engine of plate tectonics, responsible for the constant reshaping of our planet's surface. It's a slow but relentless process, measured in centimeters per year, yet over geological timescales, it leads to the formation of mountains, the opening and closing of oceans, and the distribution of volcanoes and earthquake zones.

Overlying the mantle is the Earth's crust, a thin, rocky shell. We live on this crust, and it's the most directly accessible layer for scientific investigation. The crust is divided into two main types: oceanic crust and continental crust. Oceanic crust is thinner and denser, composed primarily of basalt, a volcanic rock rich in iron and magnesium. Continental crust is thicker and less dense, consisting of a variety of rock types, with granite, a silica-rich rock, being a significant component. The contrast in density between oceanic and continental crust plays a vital role in the subduction process, where denser oceanic crust dives beneath lighter continental crust at convergent plate boundaries, contributing to volcanic activity and earthquakes. Think of it as a giant recycling system, where oceanic crust is continuously created at mid-ocean ridges and destroyed at subduction zones.

Understanding the Earth's internal structure is paramount to comprehending the processes that drive volcanic eruptions and earthquakes. The intricate interplay between the core, mantle, and crust creates the conditions necessary for these dynamic phenomena. The heat generated within the core drives mantle convection, which in turn powers the movement of tectonic plates. This movement creates stress and strain in the Earth's crust, leading to the fracturing of rocks and the release of immense energy in the form of earthquakes. Furthermore, the movement and melting of rocks in the mantle, especially near plate boundaries, leads to the formation of magma.

Magma, molten rock beneath the Earth's surface, is the lifeblood of volcanoes. Its generation is intricately linked to the dynamics of the Earth's interior. At divergent plate boundaries, the separation of plates creates a zone of reduced pressure, allowing the mantle material to rise and melt. This decompression melting is a primary mechanism for magma formation at mid-ocean ridges. At convergent plate boundaries, where oceanic crust subducts beneath continental crust, water released from the subducting slab lowers the melting point of the overlying mantle rock. This process, known as flux melting, contributes significantly to the formation of magma at subduction zones. The composition of magma varies depending on its source and the melting processes involved. Basaltic magma, derived from the mantle, is relatively low in silica content. Andesitic and rhyolitic magmas, often found in continental volcanic arcs, are richer in silica due to the involvement of continental crust in their formation. These differences in magma composition influence the explosiveness of volcanic eruptions. Silica-rich magmas tend to be more viscous, trapping gases and leading to more explosive eruptions.

The study of the Earth's interior is a continuous exploration, utilizing a variety of techniques to unravel its secrets. Seismology, the study of earthquake waves, provides valuable information about the Earth's internal structure. By analyzing the speed and paths of seismic waves as they travel through the planet, scientists can infer the density and composition of different layers. Furthermore, laboratory experiments replicating the high pressures and temperatures found deep within the Earth offer insights into the behavior of materials under these extreme conditions. These combined efforts, from seismic studies to laboratory experiments, continuously refine our understanding of the dynamic processes occurring within our planet. This knowledge is essential not only for understanding volcanoes and earthquakes but also for appreciating the interconnectedness of Earth's systems and the profound influence they have on our planet's surface and its inhabitants. The journey from the Earth's core to its crust is a testament to the immense power and delicate balance that shape our world.

1.2 Plate Tectonics

Our planet's surface, seemingly solid and unchanging, is a dynamic mosaic of colossal plates in constant, albeit incredibly slow, motion. This movement, known as plate tectonics, is the driving force behind the creation of mountains, the eruption of volcanoes, and the trembling of the earth during earthquakes. Understanding this fundamental process is key to comprehending the dynamic nature of our planet. These plates, composed of the Earth's rigid outermost layer – the lithosphere – float atop a semi-molten layer called the asthenosphere. Think of it like rafts drifting on a slow-moving river. This asthenosphere, though solid, behaves plastically over geological timescales, allowing the overlying plates to shift and interact. The heat from the Earth's core drives convection currents within the asthenosphere, creating the forces that propel these plates.

The lithosphere itself is not a uniform shell. It is fragmented into a dozen major and several minor plates, each with its own unique shape and size. These plates interact at their boundaries, and it is at these boundaries where the most dramatic geological events occur. The plates don't simply glide smoothly past each other; their interactions are complex and often violent. Friction between the plates can cause them to become locked, building up tremendous pressure over time. This stored energy is eventually released in the form of earthquakes, often with devastating consequences.

There are three primary types of plate boundaries, each characterized by a different type of plate interaction. Divergent boundaries occur where plates move apart, creating new crust as magma rises from the asthenosphere to fill the gap. This process is most evident at mid-ocean ridges, vast underwater mountain ranges that mark the sites of seafloor spreading. Convergent boundaries, on the other hand, occur where plates collide. Here, one plate typically subducts, or dives, beneath the other, descending into the mantle. This subduction process is responsible for the formation of deep ocean trenches and volcanic arcs, chains of volcanoes that parallel the convergent boundary. The third type, transform boundaries, occur where plates slide horizontally past each other. The San Andreas Fault in California is a prime example of a transform boundary, where the Pacific Plate grinds against the North American Plate.

The constant motion of these tectonic plates has profoundly shaped the Earth's surface over millions of years. Continents have drifted across the globe, oceans have opened and closed, and mountain ranges have been uplifted through the collision of plates. The Himalayas, the world's highest mountain range, are a testament to the immense power of plate tectonics, formed by the ongoing collision between the Indian and Eurasian plates. Similarly, the Andes Mountains, stretching along the western edge of South America, are the result of the Nazca Plate subducting beneath the South American Plate.

The concept of plate tectonics revolutionized our understanding of the Earth. It provides a unifying framework that explains seemingly disparate phenomena, from the distribution of volcanoes and earthquakes to the formation of mountain ranges and ocean basins. Before the development of plate tectonics theory, these processes were often viewed as isolated events. Now, we recognize them as interconnected manifestations of the Earth's dynamic internal workings. Consider the "Ring of Fire," a zone encircling the Pacific Ocean where a vast majority of the world's earthquakes and volcanoes occur. This ring directly corresponds to the boundaries of the Pacific Plate, highlighting the link between plate tectonics and these geological hazards.

Furthermore, plate tectonics plays a crucial role in the Earth's climate system. The movement of continents influences ocean currents and wind patterns, affecting global temperature distribution. The uplift of mountains due to plate collisions can alter regional climates by creating rain shadows and affecting air circulation. Even the release of volcanic gases during eruptions can impact the atmosphere, influencing short-term climate variability.

The study of plate tectonics continues to be an active area of research. Scientists are using sophisticated technologies, such as GPS and satellite imagery, to track plate movements with increasing precision. They are also developing computer models to simulate plate interactions and predict future geological events. While we have made tremendous strides in understanding plate tectonics, many questions remain. Researchers are still exploring the intricate details of how plates move and interact, the forces driving these movements, and the long-term implications for the evolution of our planet. By continuing to explore these fundamental processes, we can gain a deeper understanding of the dynamic planet we call home.

1.3 Magma Formation

Deep beneath the Earth's surface, temperatures soar, reaching thousands of degrees Celsius. This intense heat, a relic of planetary formation and fueled by radioactive decay, plays a crucial role in the creation of magma. Magma, the molten or partially molten rock that feeds volcanoes, isn't simply a uniform liquid. It's a complex mixture of molten rock, dissolved gases, and suspended crystals, varying in composition and properties depending on its origin and the conditions under which it forms. Understanding magma formation is essential to comprehending the dynamics of volcanic eruptions and the creation of various igneous rocks.

There are three primary mechanisms involved in the generation of magma: decompression melting, flux melting, and heat transfer. Decompression melting occurs when the pressure on a rock decreases without a significant change in temperature. Think of it like releasing the pressure on a shaken soda bottle; the dissolved gases expand rapidly, causing the liquid to erupt. Similarly, within the Earth, rocks in the mantle are under immense pressure. If these rocks rise towards the surface, perhaps through convection currents or along divergent plate boundaries, the decrease in pressure allows the rock to melt, even if the temperature remains relatively constant. This process is particularly significant at mid-ocean ridges, where tectonic plates diverge, and mantle rock ascends, creating vast amounts of basaltic magma.

Flux melting, on the other hand, involves the introduction of volatiles, such as water and carbon dioxide, into the mantle. These volatiles lower the melting point of the surrounding rocks, much like adding salt to ice lowers its melting point. This process is crucial in subduction zones, where the subducting oceanic plate carries water-rich sediments deep into the mantle. As the plate descends, the increasing temperature and pressure release these volatiles, triggering the melting of the overlying mantle wedge. The resulting magma, often andesitic or rhyolitic in composition, is typically more viscous and gas-rich than magma formed through decompression melting, contributing to the explosive nature of volcanic eruptions at subduction zones.

Heat transfer melting occurs when hot magma rises from the mantle and intrudes into the cooler crustal rocks. The heat from the magma conducts into the surrounding crust, raising the temperature and causing the crustal rocks to melt. This process is like placing a hot iron on a block of ice; the heat transfers from the iron to the ice, causing it to melt. This mechanism often leads to the formation of granitic magmas, which are characteristic of continental volcanic arcs and large igneous provinces. The composition of the resulting magma depends both on the composition of the intruding magma and the composition of the surrounding crustal rocks.

The composition of magma is a critical factor influencing its viscosity, or resistance to flow. Silica content plays a dominant role in determining viscosity. Magmas high in silica, such as rhyolitic magmas, tend to be very viscous and flow slowly, trapping gases and increasing the potential for explosive eruptions. In contrast, low-silica magmas, like basaltic magmas, are less viscous and flow more readily, allowing gases to escape more easily, leading to relatively gentle, effusive eruptions.

The temperature of magma also affects its viscosity. Higher temperatures reduce viscosity, making the magma flow more readily. Pressure also influences magma behavior. High pressures deep within the Earth keep volatiles dissolved within the magma. As the magma rises towards the surface, the pressure decreases, and these dissolved gases begin to form bubbles, much like the carbonation in a soda. The expansion of these bubbles can dramatically increase the pressure within the magma chamber, driving volcanic eruptions. The interplay of temperature, pressure, volatile content, and silica content governs the complex behavior of magma and the diverse nature of volcanic activity. Understanding these intricacies allows us to better interpret the signals from our planet's fiery heart, helping us to predict and mitigate the hazards associated with volcanic eruptions.

2 Chapter 2: Volcanic Eruptions

Volcanoes, conduits between the Earth's molten interior and its surface, exhibit a remarkable diversity in their eruptive styles. Understanding these variations is crucial to appreciating the powerful forces shaping our planet. One key factor influencing eruption type is magma viscosity, controlled by silica content and temperature. High-silica magmas, like rhyolite, tend to be stickier, trapping gases and leading to explosive eruptions. Low-silica magmas, like basalt, are more fluid, allowing gases to escape readily and resulting in gentler, effusive eruptions. Imagine honey slowly oozing from a jar; this mimics the behavior of basaltic lava. Contrast this with shaking a bottle of carbonated soda and suddenly releasing the cap – a much closer analogy to the explosive release of trapped gases in a rhyolitic eruption. This difference in viscosity explains why shield volcanoes, built from fluid basaltic lavas, have gentle slopes, while stratovolcanoes, formed from alternating layers of viscous lava and ash, have steep, conical profiles.

Beyond magma composition, the presence of dissolved gases plays a crucial role in eruption dynamics. These volatile components, primarily water vapor and carbon dioxide, act like the propellant in a rocket. As magma rises towards the surface, pressure decreases, allowing these dissolved gases to expand. In low-viscosity magmas, the gases can escape relatively easily, leading to lava fountains and relatively benign lava flows. In high-viscosity magmas, however, the expanding gases become trapped, building pressure until they are released explosively, fragmenting the magma into ash and other pyroclastic materials. These eruptions can be incredibly powerful, ejecting ash plumes kilometers into the atmosphere and generating devastating pyroclastic flows, scorching avalanches of hot gas and volcanic debris. Think of the difference between gently boiling a pot of water versus rapidly heating a pressure cooker – the sudden release of pressure in the latter can be catastrophic.

We can broadly categorize volcanic eruptions into several types. Hawaiian eruptions, characterized by effusive lava flows and fire fountains, represent the low-viscosity end of the spectrum. Strombolian eruptions, named after the Italian volcano Stromboli, involve moderate explosions that eject incandescent lava bombs and ash. Vulcanian eruptions, named after the island of Vulcano, are more explosive, producing dark, ash-laden eruption columns. Plinian eruptions, named after Pliny the Younger's description of the 79 AD eruption of Vesuvius, are highly explosive, sending massive plumes of ash and gas high into the stratosphere, often accompanied by devastating pyroclastic flows. These classifications, while useful, form a continuum, and eruptions can transition between types. For example, an eruption might begin with a Hawaiian phase of lava flows and then escalate to a more explosive Vulcanian or even Plinian phase as gas pressure builds.

The diversity in eruption styles results in a wide range of volcanic landforms, from gently sloping shield volcanoes to towering stratovolcanoes. Moreover, the volume and type of erupted material influence the immediate and long-term hazards posed by volcanoes. Understanding these processes is fundamental to mitigating volcanic risks. Observing precursory signs like ground deformation, gas emissions, and seismic activity allows scientists to forecast eruptions and provide timely warnings to vulnerable communities. The study of volcanic eruptions is not just an academic pursuit; it is a vital tool for protecting human lives and infrastructure in volcanically active regions.

Consider the contrasting examples of the 1980 eruption of Mount St. Helens and the ongoing eruptions of Kilauea in Hawaii. Mount St. Helens, with its viscous, gas-rich magma, produced a cataclysmic lateral blast that devastated the surrounding landscape. Kilauea, with its fluid, basaltic lava, has produced spectacular lava flows that have reshaped the coastline of the Big Island but with far less explosive force. These examples highlight the critical relationship between magma properties, eruption dynamics, and the resulting hazards. Studying these differences allows volcanologists to develop better models for predicting future eruptions and their potential impact.

Furthermore, the impact of volcanic eruptions extends beyond the immediate vicinity of the volcano. Large Plinian eruptions can inject vast quantities of ash and aerosols into the atmosphere, affecting global climate patterns. The 1815 eruption of Tambora, for instance, caused the "year without a summer," with widespread crop failures and famine due to decreased global temperatures. Therefore, understanding volcanic eruptions is crucial not only for local hazard mitigation but also for comprehending Earth's complex climate system and its history. Volcanoes, while destructive forces, are also integral to the planet's evolution, shaping landscapes, influencing climate, and contributing to the dynamic nature of our world. The study of volcanic eruptions is a multidisciplinary endeavor, drawing on geology, geophysics, geochemistry, and atmospheric science. By integrating these disciplines, we gain a more comprehensive understanding of the complex interplay of factors that govern volcanic activity. This integrated approach allows us to refine our models, improve our forecasting capabilities, and enhance our ability to protect communities from the devastating consequences of volcanic eruptions. As we continue to explore the intricate workings of volcanoes, we gain deeper insights into the powerful forces that have shaped our planet and continue to mold its future.

2.1 Types of Volcanoes

Volcanoes, while sharing the common thread of molten rock reaching the Earth's surface, exhibit a remarkable diversity in their forms, eruptive styles, and the landscapes they create. This diversity stems from differences in magma composition, gas content, and the tectonic setting in which they occur. Understanding these variations is key to appreciating the powerful forces shaping our planet. We can broadly categorize volcanoes into several main types: shield volcanoes, cinder cones, stratovolcanoes, and lava domes. Each tells a unique geological story.

Shield volcanoes are the gentle giants of the volcanic world. Characterized by their broad, gently sloping flanks resembling a warrior's shield, these volcanoes are built up primarily from successive flows of low-viscosity basaltic lava. This type of lava flows easily, spreading out over large distances before solidifying. Consequently, shield volcanoes tend to be very large in area but relatively low in height. The Hawaiian Islands, exemplified by Mauna Loa and Kilauea, are iconic examples of shield volcanoes formed by hotspot volcanism. Their eruptions are typically effusive, meaning they are characterized by the relatively quiet outpouring of lava rather than explosive blasts. Over immense periods, these lava flows build up layer upon layer, slowly constructing the massive form of the volcano.

Cinder cones, in contrast, are the smallest and simplest type of volcano. They are built from pyroclastic material—fragments of solidified lava ejected during eruptions. These fragments, ranging in size from ash to larger cinders and volcanic bombs, accumulate around the vent, forming a steep-sided conical hill. Cinder cones often form on the flanks of larger volcanoes or in volcanic fields. Their eruptions are typically short-lived and relatively explosive, driven by the rapid release of gas from the rising magma. Parícutin in Mexico, which famously erupted in a farmer's field in 1943, provides a classic example of a cinder cone's rapid formation. Its birth and growth were witnessed and documented in detail, offering valuable insights into this type of volcanic activity.

Stratovolcanoes, also known as composite volcanoes, represent a more complex and often more dangerous type of volcano. These majestic, towering peaks are formed from alternating layers of lava flows and pyroclastic deposits. The lava is typically more viscous than that of shield volcanoes, resulting in steeper slopes. The interlayering of lava and pyroclastic material reflects the varied nature of stratovolcanic eruptions, which can range from effusive lava flows to highly explosive blasts. The iconic Mount Fuji in Japan and Mount St. Helens in the United States are prime examples of stratovolcanoes. Their volatile nature arises from the trapped gases within the viscous magma, which can build up tremendous pressure leading to catastrophic eruptions. The 1980 eruption of Mount St. Helens dramatically demonstrated the destructive power of these volcanoes, reshaping the surrounding landscape in a matter of hours.

Lava domes form when highly viscous lava erupts slowly and piles up around the vent, creating a dome-shaped structure. These domes can be relatively small features on the flanks of larger volcanoes or can grow into substantial independent structures. The lava is so thick that it often does not flow far from the vent. This leads to the characteristic dome shape, sometimes with a spiny or craggy surface due to the cooling and fracturing of the viscous lava. Lava domes can be found in various volcanic settings, often associated with stratovolcanoes. Their growth can be episodic, with periods of slow extrusion punctuated by occasional collapses or explosions. The Novarupta Dome, formed during the 1912 eruption of the Katmai volcano in Alaska, is one of the largest lava domes ever documented, showcasing the scale these features can attain.

Understanding the distinct characteristics of these different volcano types is crucial for assessing volcanic hazards and mitigating their risks. The composition of the magma and the eruptive style of a volcano dictate the types of hazards posed, from slow-moving lava flows to fast-moving pyroclastic flows and widespread ashfall. By studying the morphology and eruptive history of volcanoes, scientists can gain valuable insights into the potential dangers they present and develop strategies for protecting communities living in their shadow. The science of volcanology continues to refine our understanding of these complex and dynamic systems, helping us to live more safely alongside these powerful forces of nature.

2.2 Eruption Dynamics

The violence of a volcanic eruption is a complex interplay of factors, a delicate balance between the magma's properties and the surrounding geological conditions. Magma viscosity, controlled by its silica content and temperature, plays a crucial role. A low-viscosity magma, like basalt, flows easily, leading to effusive eruptions characterized by gentle lava flows. Conversely, high-viscosity magma, richer in silica like rhyolite, resists flow, trapping gases and building pressure. This culminates in explosive eruptions, marked by dramatic explosions and pyroclastic flows. Imagine honey slowly dripping from a spoon versus shaking a carbonated drink and suddenly releasing the cap – the difference in viscosity dictates the outcome.
Furthermore, the dissolved gas content within the magma acts as the engine driving eruptions. As magma rises towards the surface, the decreasing pressure allows dissolved gases, primarily water vapor and carbon dioxide, to exsolve, forming bubbles. In low-viscosity magmas, these bubbles escape easily, contributing to the relatively peaceful effusive eruptions. However, in highly viscous magmas, the bubbles are trapped, increasing pressure within the magma chamber like an inflating balloon. This pressure buildup eventually overcomes the strength of the surrounding rock, leading to a sudden, violent release of energy. Think of it as the difference between gently opening a valve and having a pressure cooker explode – the rate of gas release dictates the eruption style.
The conduit system, or the pathway through which magma ascends, also influences eruption dynamics. The geometry and size of the conduit, as well as the presence of pre-existing fractures or weaknesses in the surrounding rock, affect the magma's ascent rate and the ease with which gases can escape. A narrow, constricted conduit can impede magma flow and increase pressure, promoting explosive eruptions. A wider, less constricted conduit allows for easier degassing and a more gentle eruption. Visualize water flowing through a narrow straw versus a wide pipe – the

diameter influences the flow rate and pressure.

External factors, such as the interaction of magma with groundwater or surface water, also contribute to the eruption dynamics. When magma comes into contact with water, it can rapidly heat the water, generating steam. This rapid expansion of steam can cause powerful phreatomagmatic eruptions, characterized by violent explosions and the ejection of ash and rock fragments. Picture a drop of water hitting a hot skillet – the instantaneous vaporization creates a miniature explosion. Similarly, the interaction of magma with external water sources significantly increases the explosivity of an eruption.

The geological setting, particularly the type of volcano, plays a critical role in shaping eruption dynamics. Shield volcanoes, with their gently sloping flanks built from low-viscosity basaltic lava flows, are typically associated with effusive eruptions. Stratovolcanoes, composed of alternating layers of lava and pyroclastic material, are prone to both effusive and explosive eruptions, reflecting the variability in magma composition and gas content. Calderas, formed by large-scale collapses following massive eruptions, can host a range of eruption styles depending on the underlying magma system. Imagine the difference between building a sandcastle with fine, wet sand versus constructing a tower with alternating layers of bricks and mortar – the material properties dictate the structure's stability and potential for collapse.

Understanding these intricate relationships between magma properties, conduit systems, external influences, and geological setting is crucial for predicting and mitigating the hazards associated with volcanic eruptions. By analyzing the composition and gas content of magma, monitoring ground deformation, and studying the history of past eruptions, scientists can assess the potential for future activity and provide valuable information for hazard assessments and evacuation planning. This is similar to a physician assessing a patient's vital signs, medical history, and lifestyle factors to diagnose and treat a potential illness – gathering and analyzing data is crucial for understanding and managing risks. Ultimately, unraveling the complexities of eruption dynamics allows us to better understand the powerful forces shaping our planet and protect vulnerable communities from volcanic hazards. This is not just an academic exercise; it's a vital endeavor that can save lives and preserve livelihoods.

3 Chapter 3: Volcanic Landforms

The Earth's surface bears the indelible mark of volcanic activity, sculpted over millennia by molten rock, ash, and gases. These processes give rise to a remarkable diversity of landforms, each telling a unique story of the forces at play beneath our feet. We begin our exploration with cinder cones, relatively small, cone-shaped structures built from ejected lava fragments. These fragments, known as scoria or cinders, cool rapidly in the air and accumulate around the vent, forming a steep-sided cone. Cinder cones often occur in groups and are typically associated with basaltic eruptions, a type of lava that is relatively fluid. Their slopes are generally unstable and prone to erosion, reflecting the loose nature of the accumulated material. In contrast to the dramatic peaks of stratovolcanoes, cinder cones represent a more localized, yet equally fascinating, expression of volcanic power.

Shield volcanoes, on the other hand, are built by the accumulation of countless low-viscosity lava flows. These flows spread out over vast distances, creating broad, gently sloping domes that resemble a warrior's shield, hence their name. Mauna Loa and Kilauea in Hawaii are quintessential examples of shield volcanoes, rising gradually from the ocean floor to form imposing landmasses. The lava emitted from shield volcanoes is typically basaltic, allowing it to travel considerable distances before solidifying. Eruptions are generally effusive, characterized by steady lava flows rather than explosive bursts, contributing to their characteristic shape. The sheer size of shield volcanoes testifies to the immense volume of lava erupted over prolonged periods.

Stratovolcanoes, also known as composite volcanoes, present a more dramatic profile. Their steep, conical shapes are constructed from alternating layers of lava flows, volcanic ash, and other pyroclastic materials. This layered structure reflects the cyclical nature of their eruptions, alternating between effusive and explosive phases. Mount Fuji in Japan and Mount St. Helens in the United States are iconic examples of stratovolcanoes. The viscous, silica-rich lava associated with stratovolcanoes tends to solidify quickly, leading to the build-up of pressure and ultimately explosive eruptions. These eruptions can be highly destructive, ejecting vast quantities of ash and pyroclastic flows that pose significant hazards to surrounding areas.

Calderas represent a different kind of volcanic landform, formed by the collapse of a volcano's summit following a massive eruption that empties the underlying magma chamber. This collapse creates a large, basin-shaped depression, often many kilometers in diameter. Crater Lake in Oregon, nestled within a caldera, is a striking example of this phenomenon. Calderas can also form through the gradual subsidence of a magma chamber without a catastrophic eruption. These diverse processes highlight the dynamic interplay between internal magma movements and surface topography. The formation of a caldera signifies a major event in a volcano's lifespan, dramatically altering the landscape and often leaving behind a lingering reminder of the powerful forces at work.

Lava domes, often found within or near volcanic craters or calderas, are formed by the extrusion of highly viscous lava. This viscous lava, typically rhyolitic or dacitic in composition, resists flowing and instead piles up around the vent, creating a dome-shaped structure. These domes can grow slowly over time or experience rapid growth spurts. Their steep sides and unstable nature make them susceptible to collapse, potentially generating pyroclastic flows. The slow extrusion of lava in dome formation contrasts sharply with the rapid, flowing eruptions characteristic of shield volcanoes.

Beyond individual volcanic structures, vast expanses of the Earth's surface are covered by lava flows and plateaus. These features result from the accumulation of numerous lava flows over extended periods, covering extensive areas and creating relatively flat or gently sloping terrain. The Deccan Traps in India and the Columbia River Basalt Group in the northwestern United States are prime examples of flood basalt provinces, demonstrating the enormous scale of these outpourings. These vast lava plateaus offer insights into periods of intense volcanic activity in Earth's history.

Volcanism isn't confined to the land; it also occurs beneath the ocean's surface. Submarine volcanoes, often found along mid-ocean ridges, erupt basaltic lava that quickly cools in contact with seawater, forming pillow-like structures. Hydrothermal vents associated with these underwater volcanoes support unique ecosystems, thriving in the chemically rich, superheated water. Submarine eruptions contribute significantly to the formation of new oceanic crust, adding to our understanding of plate tectonics and the dynamic nature of the seafloor. These hidden volcanic landscapes play a crucial role in shaping the ocean floor and sustaining unique biological communities. The diverse range of volcanic landforms, from the subtle slopes of shield volcanoes to the dramatic peaks of stratovolcanoes and the hidden depths of submarine eruptions, paints a vivid picture of Earth's dynamic nature. These features not only shape our planet's surface but also offer valuable clues about the processes occurring deep within its interior.

3.1 Cinder Cones & Shields

Imagine standing on the flanks of a volcano, its slopes a uniform, gentle incline stretching out as far as the eye can see. This broad, dome-like structure is a shield volcano, its form a testament to the low-viscosity lava that has flowed from its vents over millennia. Hawaiian volcanoes like Mauna Loa and Kilauea exemplify this type. Their lava, often basaltic, is relatively fluid, allowing it to travel great distances before solidifying. This creates the characteristically broad, shield-like profile, often resembling a warrior's shield laid upon the land. These eruptions are typically effusive, not explosive, characterized by the relatively calm outpouring of lava. The molten rock, rich in iron and magnesium, flows like rivers of fire, gradually building up the volcano layer by layer. This slow accumulation of lava can create immense structures – Mauna Loa, measured from its base on the ocean floor, is the largest mountain on Earth.

Contrast this with a cinder cone. These are smaller, steeper volcanic features built from fragmented volcanic rock called tephra. Picture a conical hill, its slopes composed of loose, dark cinders and ash. Unlike the steady flow of shield volcanoes, cinder cones are born from more volatile eruptions. Gas-rich magma explodes skyward, showering the surrounding landscape with a rain of solidified lava fragments. These fragments, ranging in size from fine ash to larger cinders and volcanic bombs, accumulate around the vent, building up the cone's steep slopes. Cinder cones often rise quickly, sometimes forming in a single eruptive episode, but they are also relatively short-lived. Their unconsolidated nature makes them prone to erosion, and their smaller size means they are often dwarfed by their larger volcanic cousins. Parícutin in Mexico, a cinder cone that emerged from a farmer's field in 1943, famously grew to several hundred meters in just a few years.

The composition of the magma plays a crucial role in determining whether a volcano forms a shield or a cinder cone. Shield volcanoes are associated with basaltic magma, low in silica content and thus less viscous. This allows the lava to flow readily, spreading out over wide areas. The Hawaiian Islands, formed over a hotspot where basaltic magma rises from deep within the Earth's mantle, are a prime example of shield volcano formation. In contrast, cinder cones often form from magma with a higher silica content, making it stickier and more resistant to flow. This viscous magma traps gases, building up pressure that eventually leads to explosive eruptions. The resulting fragments are then ejected to form the characteristic cone. Sunset Crater in Arizona, a vivid red cinder cone, exemplifies this type of volcanism.

The internal structure of these volcanoes also differs significantly. Shield volcanoes possess a network of interconnected lava tubes, underground conduits that allow molten rock to flow far from the central vent. These tubes act like insulated pipes, keeping the lava hot and fluid, enabling it to travel great distances before cooling. This contributes to the volcano's broad, gently sloping profile. Cinder cones, on the other hand, have a simpler internal structure. They typically consist of a central vent filled with loose tephra, with no extensive lava tube system. The fragments accumulate around the vent, forming the cone's steep, unstable slopes. This difference in internal structure reflects the contrasting eruptive styles of these two volcano types.

The landscapes surrounding cinder cones and shield volcanoes further reflect their distinct characteristics. Shield volcanoes are often associated with vast lava fields, solidified rivers of basalt that stretch for miles. These fields can be barren and desolate, or they can support unique ecosystems adapted to the volcanic environment. Cinder cones, due to their smaller size and more localized eruptions, often create a more contained impact. The surrounding landscape might be littered with tephra, but the area affected is typically smaller than that impacted by shield volcano eruptions. The loose, porous tephra can also influence local hydrology, affecting groundwater infiltration and runoff patterns. Moreover, the rich mineral content of volcanic ash can contribute to fertile soils, supporting the eventual return of vegetation.

While shield volcanoes and cinder cones represent two distinct volcanic landforms, they can sometimes coexist. In some volcanic regions, cinder cones may form on the flanks of a larger shield volcano. This can occur when eruptive activity shifts from the central vent to a new location on the volcano's slope. The resulting cinder cone becomes a parasitic cone, drawing its magma from the same source as the shield volcano. Such combinations demonstrate the complex interplay of factors that shape volcanic landscapes. The study of these different volcanic forms provides valuable insights into the inner workings of our planet and the powerful forces that shape its surface. Understanding these processes is crucial not only for scientific knowledge but also for mitigating the risks associated with volcanic hazards.

3.2 Stratovolcanoes

Majestic and imposing, stratovolcanoes represent some of the most iconic and hazardous volcanic landforms on Earth. Their steep, symmetrical cones, often snow-capped, dominate skylines and inspire both awe and apprehension. This distinctive shape is a direct result of their formation process, a complex interplay of eruptive styles and material composition. Unlike the gently sloping shields formed by fluid basaltic lava or the simple cones of scoria cinders, stratovolcanoes are constructed layer upon layer, hence their name, "strato" meaning layered. These layers, or strata, consist of alternating deposits of pyroclastic material—ash, pumice, and rock fragments—and solidified lava flows. This interbedding gives stratovolcanoes their strength and structural integrity, allowing them to reach towering heights.

The viscous nature of the magma feeding stratovolcanoes plays a critical role in their characteristically explosive eruptions. This magma, typically andesitic or dacitic in composition, is richer in silica than basaltic magma, making it thicker and less likely to flow easily. As the magma rises through the conduit system, trapped gases, including water vapor and carbon dioxide, exert increasing pressure. This pressure builds until it overcomes the strength of the surrounding rock, resulting in a violent release of pent-up energy. Eruptions can range from relatively small, intermittent events to cataclysmic explosions that eject vast quantities of ash and debris into the atmosphere. Mount St. Helens' 1980 eruption serves as a stark reminder of the destructive potential of these volcanoes, demonstrating the devastating power of pyroclastic flows and lahars.

The eruptive history of a stratovolcano often reveals a pattern of alternating effusive and explosive activity. Periods of relatively quiet lava flows build up the volcano's flanks, adding to its overall size and stability. These lava flows, while still viscous, can travel considerable distances down the volcano's slopes, solidifying to form robust, resistant layers. However, these periods of quiescence can be punctuated by sudden, violent explosive eruptions. These eruptions eject massive amounts of pyroclastic material, which can blanket the surrounding landscape in ash and pumice. The rapid deposition of this material creates steep slopes, further contributing to the classic conical profile of a stratovolcano. This cyclical process of lava flow construction and explosive deconstruction is what ultimately shapes the volcano's final form.

The location of stratovolcanoes is intimately linked to plate tectonics. They are predominantly found along convergent plate boundaries, specifically in subduction zones, where one tectonic plate slides beneath another. As the subducting plate descends into the Earth's mantle, it releases water and other volatiles into the overlying mantle wedge. This influx of volatiles lowers the melting point of the mantle rock, generating magma. This magma, now buoyant, begins its ascent through the overlying crust. As it rises, it undergoes a process of fractional crystallization, whereby certain minerals crystallize and settle out, altering the magma's composition and increasing its viscosity. This process further contributes to the explosive nature of stratovolcanic eruptions. The "Ring of Fire" encircling the Pacific Ocean, a zone of intense volcanic and seismic activity, provides a striking illustration of this connection, hosting a significant number of the world's stratovolcanoes.

Beyond the immediate vicinity of the volcano, stratovolcanic eruptions can have far-reaching consequences. The ejection of vast quantities of ash and aerosols into the atmosphere can disrupt air travel and impact global climate patterns. Fine ash particles can remain suspended in the stratosphere for extended periods, reflecting sunlight back into space and causing a temporary cooling effect. The 1815 eruption of Mount Tambora in Indonesia, a stratovolcano, led to the "Year Without a Summer" in 1816, characterized by widespread crop failures and food shortages in the Northern Hemisphere. Studying these past events allows scientists to better understand the potential impacts of future eruptions and develop strategies for mitigating their effects.

Understanding the processes that govern the formation and behavior of stratovolcanoes is crucial for assessing and mitigating volcanic hazards. Monitoring techniques, including seismic monitoring, gas measurements, and ground deformation studies, provide valuable insights into the internal workings of these complex systems. By carefully analyzing these data, scientists can identify signs of unrest and potentially forecast future eruptions, providing crucial warnings to communities living in the shadow of these magnificent yet dangerous giants. While the beauty and power of stratovolcanoes are undeniable, it is through scientific understanding that we can learn to coexist with these formidable forces of nature.

3.3 Calderas & Domes

Imagine a landscape scarred by volcanic activity. Among the varied formations, two stand out: calderas and domes. These structures, though distinct in their formation and appearance, often coexist, telling a complex story of volcanic evolution. Their presence offers invaluable clues to the history of a volcano and the forces shaping the Earth's surface. This chapter explores the formation, characteristics, and significance of calderas and domes, demonstrating how they contribute to our understanding of volcanic processes.

Calderas, immense basin-shaped depressions, represent dramatic collapses in the volcanic landscape. Their formation begins with a massive eruption that empties a substantial portion of the underlying magma chamber. The unsupported roof of the chamber then collapses inward, creating a vast, often circular depression many times larger than the original vent. Think of it like a deflated balloon, its skin collapsing once the air is removed. The size of a caldera can be staggering, ranging from a few kilometers to tens of kilometers in diameter. Notable examples include Yellowstone caldera in Wyoming, a colossal structure formed by multiple eruptions over millions of years, and Crater Lake in Oregon, a stunning example of a caldera filled with water. The scale of these features underscores the immense power of the volcanic events that create them.

However, not all calderas result from single cataclysmic eruptions. Some form gradually, through a series of smaller eruptions that progressively drain the magma chamber, causing a piecemeal collapse. This process, known as piecemeal subsidence, often results in a complex caldera structure with multiple nested depressions. Regardless of their formation mechanism, calderas signify a significant stage in a volcano's life cycle, often marking a transition to a less active phase, although renewed activity within a caldera is certainly possible.

In contrast to the concave form of a caldera, volcanic domes rise above the surrounding terrain. These steep-sided mounds form when viscous, silica-rich lava erupts slowly and piles up around the vent. The thick lava doesn't flow readily, like the more fluid basaltic lavas that create shield volcanoes. Instead, it extrudes like toothpaste from a tube, building a dome-shaped structure that can reach hundreds of meters in height. The shape and size of a dome depend on several factors, including the viscosity of the lava, the rate of eruption, and the underlying topography.

The lava that forms domes is often associated with explosive eruptions due to its high gas content. As the lava rises, the pressure decreases, allowing dissolved gases to expand. However, the viscous nature of the lava traps these gases, building pressure until it is released in explosive bursts. This can create a hazardous situation, as the dome can become unstable and collapse, triggering pyroclastic flows, which are fast-moving currents of hot gas and volcanic debris. The 1902 eruption of Mount Pelée on the island of Martinique provides a tragic example of the devastating power of dome collapses and associated pyroclastic flows.

The relationship between calderas and domes can be complex and intertwined. Domes often form within calderas, representing renewed volcanic activity following a caldera-forming eruption. This occurs as magma slowly recharges the depleted magma chamber and finds its way to the surface. The viscous lava then erupts to form a dome within the caldera's depression. This relationship is evident in many caldera systems around the world. For example, within the vast caldera of Santorini in Greece, several volcanic domes have emerged, creating a striking landscape of contrasting forms. These domes represent the volcano's continued activity, albeit at a lower intensity compared to the caldera-forming eruption that shaped the island.

Studying calderas and domes provides crucial insights into volcanic processes. The size and shape of a caldera can reveal the magnitude of the eruption that formed it and the volume of the underlying magma chamber. The composition of the lava forming the dome offers clues to the magma's origin and the chemical processes occurring beneath the surface. Additionally, the presence of multiple domes within a caldera, along with their ages and compositions, can help reconstruct the volcano's history and identify patterns of activity.

Understanding these structures is not merely an academic exercise. It is essential for assessing volcanic hazards and mitigating their impact on surrounding communities. Calderas and domes can pose significant threats. Calderas, despite their dormant appearance, can be the sites of future eruptions. Domes, prone to instability, can collapse and generate deadly pyroclastic flows. By studying the geological features of calderas and domes, scientists can better anticipate potential hazards and develop strategies to protect vulnerable populations. This knowledge is fundamental for effective volcanic hazard management and informs land-use planning and emergency response protocols. The study of these volcanic features thus contributes not only to our understanding of the Earth's dynamic processes but also to the safety and well-being of communities living in the shadow of volcanoes.

3.4 Lava Flows & Plateaus

Molten rock, escaping the confines of the Earth's interior, doesn't always explode in dramatic, mountain-building eruptions. Sometimes, it oozes forth, flowing like a fiery river across the landscape. These lava flows, depending on their viscosity and the surrounding terrain, can sculpt a variety of distinctive features. Understanding the characteristics of these flows allows us to interpret the geological history of volcanic regions and appreciate the powerful forces that shape our planet. One key factor influencing the behavior of a lava flow is its silica content. High-silica lava, like rhyolite, tends to be thick and viscous, moving sluggishly and forming steep-sided mounds or domes. Low-silica lava, like basalt, is much more fluid. It can travel great distances, forming thin, widespread sheets. Imagine honey slowly dripping from a spoon versus water quickly spreading across a tabletop. This analogy helps visualize the dramatic difference in flow behavior.

Consider pahoehoe lava, a type of basaltic flow characterized by its smooth, ropy surface. This texture results from the cooling and solidifying of the flow's outer skin while molten lava continues to move beneath. The skin wrinkles and folds, creating the distinctive rope-like patterns. Contrast this with a'a lava, another basaltic flow that has a rough, jagged, and clinkery surface. A'a flows are also hotter and move faster than pahoehoe, and the constant breaking and tumbling of the solidified crust creates the fragmented surface. These different types of lava flows tell a story about the eruption conditions and the composition of the magma. Examining their features helps geologists decipher the volcanic past.

Beyond individual flows, repeated eruptions can build up vast, elevated regions known as lava plateaus. These plateaus form when highly fluid basalt lava erupts from fissures, rather than a central vent. The lava spreads out over large areas, layer upon layer, eventually creating an extensive, flat-topped plateau. The Columbia River Basalt Group, covering parts of Washington, Oregon, and Idaho, is a prime example of such a plateau. It represents a massive outpouring of lava over millions of years, demonstrating the immense scale of these geological processes. Think of these plateaus as giant, layered cakes, each layer representing a separate lava flow. Their vastness testifies to the Earth's impressive capacity to generate and release molten rock.

Delving into specific examples, consider the Deccan Traps in India, one of the largest volcanic provinces on Earth. Formed over 66 million years ago, these flood basalts cover an area of almost 500,000 square kilometers. Their formation coincided with the Cretaceous-Paleogene extinction event, suggesting a possible link between massive volcanic eruptions and global environmental change. This potential connection highlights the significant impact volcanic activity can have on the planet's history and evolution. The Deccan Traps offer a glimpse into a dramatic period in Earth's past, reminding us of the power and scale of volcanic processes.

Another fascinating example is the Snake River Plain in Idaho. This plain represents a track of volcanic activity related to the movement of the North American plate over a hotspot. Over millions of years, as the plate moved westward, a chain of volcanic centers formed, leaving a trail of calderas, lava flows, and other volcanic features. The Snake River Plain offers a unique window into the dynamics of hotspots and their interaction with tectonic plates. This region showcases the interplay between volcanic activity and plate tectonics, illustrating the continuous reshaping of Earth's surface.

Finally, let's consider the unique environment of submarine volcanism. Eruptions occurring beneath the ocean's surface create distinct landforms and contribute to the chemical composition of seawater. Pillow lava, characterized by its rounded, pillow-like shapes, is a common product of underwater eruptions. The rapid cooling of lava in contact with seawater forms a solidified crust, while molten lava continues to flow inside, creating the characteristic bulbous forms. Pillow lava provides evidence of past submarine volcanic activity and offers insights into the interactions between molten rock and water. These underwater eruptions, hidden from our direct view, play a significant role in shaping the ocean floor and influencing the global environment. They demonstrate the ongoing, dynamic processes occurring beneath the waves.

3.5 Submarine Volcanism

The ocean floor, hidden beneath kilometers of water, is far from a static, serene environment. It's a dynamic landscape sculpted by immense geological forces, one of the most significant being submarine volcanism. This underwater volcanic activity is responsible for creating a vast array of geological features, from sprawling seamounts and volcanic ridges to hydrothermal vent fields teeming with unique life forms. Unlike their terrestrial counterparts, submarine volcanoes operate under immense pressure, which profoundly affects the nature of their eruptions. Water's high heat capacity also plays a critical role, rapidly cooling the ejected lava and generating large quantities of steam. This interaction between molten rock and seawater creates distinctive formations like pillow lavas, formed when lava erupts onto the seafloor and rapidly cools, forming rounded, pillow-like structures. These formations are a hallmark of submarine eruptions and are rarely found on land.

The majority of submarine volcanoes are located along mid-ocean ridges, vast underwater mountain ranges that mark the boundaries of diverging tectonic plates. As plates move apart, magma rises from the mantle to fill the gap, creating new oceanic crust in a process known as seafloor spreading. This continuous volcanic activity contributes significantly to the chemical composition of the ocean. The release of gases and minerals from hydrothermal vents associated with these ridges influences ocean salinity and provides nutrients for specialized ecosystems. These vent systems, often referred to as "black smokers" due to the dark, mineral-rich plumes they emit, support unique communities of organisms that thrive on chemosynthesis rather than photosynthesis. They represent an extraordinary example of life flourishing in an environment once thought to be

completely inhospitable.
Away from mid-ocean ridges, submarine volcanoes can also arise from hotspots, plumes of magma originating deep within the Earth's mantle. These hotspots remain stationary while tectonic plates drift over them, leaving a trail of volcanic islands and seamounts like the Hawaiian Islands. The composition of magma from hotspot volcanoes differs from that of mid-ocean ridge volcanoes, providing valuable insights into the chemical heterogeneity of the Earth's mantle. Analyzing the isotopic ratios of elements within these lavas can help geologists trace the origin and evolution of mantle plumes. This information contributes to our understanding of the large-scale processes that shape our planet.
Submarine eruptions, often occurring at great depths, can be difficult to detect and study. However, advanced technologies like remotely operated vehicles (ROVs) and sonar mapping have enabled scientists to explore these underwater volcanic landscapes in unprecedented detail. These tools provide crucial information about the morphology and distribution of submarine volcanoes, as well as the dynamics of their eruptions. Hydroacoustic monitoring can detect the subtle pressure changes and acoustic signals associated with underwater volcanic activity, providing early warnings of potential eruptions. While direct observation remains challenging, these advancements allow for a more comprehensive understanding of the processes shaping the ocean floor.
The study of submarine volcanism is not only crucial for understanding the geological evolution of the Earth, but also for assessing potential hazards. Although most submarine eruptions pose no direct threat to human populations, larger eruptions near the surface can trigger tsunamis, devastating waves that can travel thousands of kilometers across the ocean. The Krakatoa eruption of 1883, a primarily submarine eruption, generated a massive tsunami that claimed tens of thousands of lives. Understanding the dynamics of submarine eruptions and their potential to generate tsunamis is critical for developing effective warning systems and mitigation strategies. Furthermore, the release of volcanic gases, particularly

sulfur dioxide, can impact the Earth's climate by altering atmospheric chemistry and affecting global temperatures.

Beyond the potential hazards, submarine volcanoes play a crucial role in the Earth's biogeochemical cycles. They contribute significantly to the chemical balance of the oceans, influencing the distribution of elements and providing nutrients that support unique ecosystems. The hydrothermal vents associated with these volcanoes are home to a remarkable diversity of life, including extremophile organisms that thrive in extreme temperatures and chemical conditions. Studying these ecosystems provides valuable insights into the limits of life on Earth and may even offer clues about the potential for life on other planets. The ongoing exploration of submarine volcanism continues to reveal fascinating details about our planet's dynamic interior and its interconnectedness with the oceans. This hidden world, once inaccessible and largely unknown, is gradually yielding its secrets, enriching our understanding of Earth's complex systems and offering valuable insights into the forces that shape our world.

4 Chapter 4: Volcanic Hazards

Volcanic eruptions, while awe-inspiring displays of nature's power, pose significant threats to human populations and their surroundings. Understanding these hazards is the first step towards mitigating their devastating impacts. This chapter will explore the diverse dangers associated with volcanic activity, focusing on the direct threats posed by lava flows and pyroclastic flows, as well as the secondary hazards of lahars and volcanic gases. Each of these phenomena presents unique challenges and requires specific mitigation strategies. We'll examine the characteristics of each hazard, their potential impact, and methods for minimizing risk.

Lava flows, molten rock that pours from volcanic vents, can incinerate everything in their path. Their speed and viscosity vary depending on the lava's composition, temperature, and the slope of the terrain. While generally slow-moving, allowing people time to evacuate, lava flows can destroy infrastructure, burying homes and agricultural lands under a thick layer of solidified rock. Predicting the path of lava flows remains a challenge, as they can be influenced by subtle changes in topography. Diversion efforts, including building barriers and cooling the lava with water, have met with limited success, highlighting the destructive power of these molten rivers. Monitoring and early warning systems are crucial for minimizing the impact on communities located near active volcanoes. Understanding the specific geological context of a volcano and its past eruptive history is essential for effective hazard assessment and planning.

Pyroclastic flows, arguably the most dangerous volcanic hazard, are turbulent avalanches of hot gas, ash, and rock fragments that race down volcanic slopes at incredible speeds, often exceeding hundreds of kilometers per hour. These scorching clouds can incinerate anything in their path, leaving little chance for escape. The immense heat and velocity of pyroclastic flows make them incredibly destructive, capable of flattening forests, burying towns, and causing widespread devastation. The eruption of Mount Vesuvius in 79 AD, which buried Pompeii and Herculaneum, serves as a chilling reminder of the destructive power of these flows. Mapping past flow deposits and monitoring volcanic activity for signs of impending eruptions are critical for mitigating the risk posed by pyroclastic flows. Effective evacuation plans are essential in areas prone to these deadly events.

Beyond the immediate dangers of lava and pyroclastic flows, volcanoes also present secondary hazards, like lahars, which can be equally devastating. Lahars are volcanic mudflows composed of water, ash, and rock fragments that flow down volcanic slopes, often following river valleys. They can be triggered by heavy rainfall on loose volcanic deposits, the melting of glaciers or snow during an eruption, or the breakout of a crater lake. Lahars can travel long distances, carrying debris and burying valleys under thick layers of mud and rock. Their speed and density make them extremely destructive, capable of sweeping away bridges, buildings, and entire communities. The 1985 Nevado del Ruiz eruption in Colombia tragically demonstrated the devastating power of lahars when a relatively small eruption triggered massive mudflows that buried the town of Armero, killing over 23,000 people. Hazard mapping, early warning systems, and community education are vital for mitigating the risk posed by lahars.

Volcanic gases, though often less visible than other hazards, can pose significant threats to human health and the environment. Eruptions release a variety of gases, including sulfur dioxide, carbon dioxide, hydrogen sulfide, and water vapor. Sulfur dioxide can react with water vapor in the atmosphere to form sulfuric acid aerosols, which can contribute to acid rain and affect air quality over large areas. High concentrations of carbon dioxide can be asphyxiating, while hydrogen sulfide is a highly toxic gas. Volcanic gases can also contribute to climate change by influencing the Earth's radiative balance. Monitoring gas emissions, understanding their dispersion patterns, and establishing air quality standards are essential for protecting public health and the environment. The long-term effects of volcanic gases on regional and global climate remain an active area of research. Scientists continue to study the complex interactions between volcanic emissions and atmospheric processes to better understand their potential impact on our planet. This research is crucial for developing effective strategies to mitigate the potential risks associated with volcanic gases.

4.1 Lava & Pyroclastic Flows

Molten rock, reaching temperatures exceeding 1,000°C, doesn't simply ooze harmlessly from a volcanic vent. It surges forth, a fiery river of destruction, as a lava flow. Understanding the behavior of these flows is critical for assessing volcanic hazards. Lava's viscosity, influenced by its silica content and temperature, dictates its speed and reach. High-silica lava, like rhyolite, creates thick, slow-moving flows that rarely travel far, often forming steep-sided domes near the vent. Conversely, low silica basalt, with its lower viscosity, can flow rapidly for kilometers, spreading out in thin sheets that can inundate vast areas. These basaltic flows can pose a significant threat to infrastructure and agriculture, though they are less immediately dangerous to human life due to their slower speeds. The surface of a lava flow often cools and solidifies, creating an insulating crust that allows the molten interior to continue flowing, extending the reach and duration of the flow's destructive path. Examining the composition and temperature of emerging lava allows volcanologists to predict its likely behavior, assisting with evacuation plans and hazard mitigation.

Beyond the immediate threat of incineration, lava flows can cause widespread damage through secondary effects. The intense heat can ignite wildfires, consuming vegetation and structures far beyond the flow's edge. When lava encounters water, it can generate explosive steam explosions, sending scalding debris and hot rocks hurtling through the air. Even slow-moving flows can overwhelm and bury infrastructure, including roads, bridges, and buildings, disrupting transportation and displacing communities for extended periods. The Hawaiian Islands, born from basaltic shield volcanoes, offer a clear example of both the constructive and destructive power of lava flows. While the flows create new land, they also frequently threaten homes and communities situated on the volcano's slopes.

Distinct from lava flows, pyroclastic flows represent an even more devastating volcanic hazard. These are turbulent avalanches of hot gas, ash, and rock fragments that cascade down the flanks of a volcano at speeds exceeding 100 kilometers per hour. Generated by the collapse of an eruption column or the explosive destruction of a volcanic dome, pyroclastic flows can incinerate everything in their path. The extreme temperatures, sometimes reaching 700°C, coupled with the high speed and density of the flow, make survival virtually impossible for anyone caught in its path. The sheer destructive force of a pyroclastic flow is capable of leveling forests, burying towns, and instantly vaporizing any water it encounters, creating a ghostly steam trail in its wake.

The infamous eruption of Mount Vesuvius in 79 AD, which buried the Roman cities of Pompeii and Herculaneum, serves as a chilling testament to the power of pyroclastic flows. The remains of the cities, frozen in time by the volcanic debris, reveal the instantaneous and catastrophic nature of the event. The city of Herculaneum, located closer to the volcano, was initially spared from the ashfall that blanketed Pompeii. However, it was ultimately engulfed by a series of pyroclastic surges, which carbonized organic matter, including wood, food, and even the inhabitants themselves, leaving behind haunting skeletal remains. The different types of pyroclastic flows, from dense, ground-hugging flows called block-and-ash flows to more dilute, turbulent surges, each present unique hazards based on their speed, temperature, and composition.

Understanding the specific characteristics of a volcano and its past eruptive history is crucial for forecasting the likelihood of pyroclastic flow generation. Scientists use a variety of monitoring techniques, including seismic monitoring, gas emissions measurements, and ground deformation studies, to detect signs of unrest that may precede a pyroclastic flow. Furthermore, detailed mapping of past flow deposits helps define hazard zones, informing land-use planning and evacuation procedures. While predicting the precise timing and magnitude of a pyroclastic flow remains a challenge, these scientific endeavors provide vital tools for mitigating the risk posed by these devastating volcanic phenomena. Recognizing the subtle signs of an impending eruption, such as changes in gas emissions or ground swelling, can provide valuable time for communities to prepare and evacuate, potentially saving countless lives.

The destructive power of both lava flows and pyroclastic flows underscores the importance of respecting the immense forces at play within our planet. By studying these phenomena and developing effective monitoring and mitigation strategies, we can strive to coexist with volcanoes, appreciating their role in shaping our world while minimizing the risks they pose to human populations. The continuing advancements in volcanology offer hope for improved forecasting and preparedness, paving the way for safer communities in volcanically active regions. These efforts are not simply about predicting eruptions, but about understanding the complex interplay of geological processes that govern these powerful natural events.

4.2 Lahars & Gas

Volcanic eruptions are not just spectacular displays of Earth's power; they are also significant natural hazards. While lava flows and pyroclastic surges pose immediate and visible threats, the insidious nature of lahars and volcanic gases often goes underestimated. These two, seemingly disparate phenomena, represent significant dangers, impacting areas far beyond the immediate vicinity of a volcano and posing unique challenges for both scientists and surrounding communities. I have witnessed the devastating effects of these volcanic hazards firsthand, and I can tell you, their power should never be underestimated.

Lahars, sometimes called volcanic mudflows or debris flows, are rapidly flowing mixtures of volcanic debris and water. They resemble rivers of wet concrete, possessing the power to carry everything from boulders the size of cars to entire uprooted trees. Their formation can be triggered by several mechanisms. Melting glaciers and snowpack during an eruption are particularly concerning, providing copious amounts of water to mobilize volcanic ash and debris on the volcano's slopes. Heavy rainfall on unconsolidated volcanic deposits can have a similar effect, even years after an eruption. Sometimes, the sudden emptying of a crater lake, whether through overflow or breaching, can unleash a catastrophic lahar. These flows can travel tens of kilometers from their source, burying valleys and communities under thick layers of debris. The 1985 eruption of Nevado del Ruiz in Colombia tragically demonstrated the destructive potential of lahars, where a relatively small eruption melted glacial ice, triggering lahars that entombed the town of Armero, killing over 23,000 people. This event underscores the importance of understanding a volcano's history, topography, and the potential sources of water in its vicinity to predict and mitigate the risk of lahars.

Volcanic gases, while often invisible, pose a significant threat, both locally and globally. During an eruption, a complex mixture of gases is released, including water vapor, carbon dioxide, sulfur dioxide, hydrogen sulfide, and various halogens. Some, like sulfur dioxide, react with water vapor in the atmosphere to form aerosols, which can reflect sunlight and cause short-term cooling of the planet, as observed after the 1991 eruption of Mount Pinatubo. Carbon dioxide, a greenhouse gas, can contribute to long-term climate change when released in substantial amounts over geological timescales. However, the most immediate danger from volcanic gases lies in their toxicity. High concentrations of sulfur dioxide and hydrogen sulfide can cause respiratory problems and even asphyxiation, particularly in low-lying areas where these denser-than-air gases can accumulate. The silent nature of these gases makes them particularly insidious, requiring specialized monitoring equipment for detection. Furthermore, the wind can carry volcanic gases far from their source, affecting communities miles away.

Predicting and mitigating the hazards of lahars and volcanic gases involves a multi-faceted approach. Detailed mapping of volcanic deposits and river valleys helps identify areas at risk from lahars. Early warning systems, using seismic monitoring and ground deformation measurements, can detect volcanic unrest and potential lahar triggers. These systems, coupled with community education and evacuation plans, are crucial for minimizing casualties. For volcanic gases, continuous monitoring of gas emissions helps assess the potential for hazardous concentrations. Scientists use a variety of instruments, including spectrometers and gas sensors, to measure the composition and concentration of volcanic gases. This data provides crucial insights into the state of the volcano and potential eruption hazards. Furthermore, understanding the dispersion patterns of volcanic gases based on wind direction and topography is essential for predicting which communities may be affected.

Furthermore, the long-term impacts of volcanic gases warrant careful consideration. While large eruptions can cause temporary global cooling, the continuous release of volcanic gases throughout Earth's history has played a crucial role in shaping our atmosphere and climate. Researchers study the composition of volcanic gases to better understand the processes occurring deep within the Earth and the long-term interactions between volcanoes and the global environment. The study of volcanic gases also provides valuable insights into the formation of early planetary atmospheres and the potential for life on other planets. By analyzing the composition of gases emitted from volcanoes on other celestial bodies, scientists can infer the presence of specific elements and compounds, potentially indicating the presence of liquid water or even biological activity.

In conclusion, understanding the nature and behavior of lahars and volcanic gases is vital for mitigating their potential impacts. These hazards, often overlooked in the face of more dramatic volcanic phenomena, pose significant threats to human populations and the environment. By combining scientific monitoring with community preparedness, we can strive to minimize the risks associated with these silent but deadly volcanic phenomena. Continued research into the composition, behavior, and long-term impacts of these volcanic hazards is crucial for protecting vulnerable communities and advancing our understanding of Earth's dynamic systems.

5 Chapter 5: Earthquake Basics

Earthquakes are fundamentally caused by the movement of Earth's tectonic plates. These plates, vast pieces of the Earth's lithosphere, are in constant, albeit slow, motion. When these plates interact, tremendous stress builds up along their boundaries, particularly along faults. Faults are fractures in the Earth's crust where rocks on either side have moved past each other. This accumulated stress eventually overcomes the friction holding the rocks together, resulting in a sudden release of energy. This energy radiates outward from the point of rupture, known as the hypocenter or focus, in the form of seismic waves. These waves travel through the Earth, causing the ground to shake. Understanding faults and seismic waves is crucial to grasping the mechanics of earthquakes.

Seismic waves, the vibrations generated by earthquakes, come in various types. The primary waves, or P-waves, are compressional waves, similar to sound waves, that travel the fastest. They can move through solids, liquids, and gases, causing the ground to move back and forth in the direction the wave is traveling. Secondary waves, or S-waves, are shear waves that travel slower than P-waves and can only propagate through solids. S-waves cause the ground to move perpendicular to the direction of wave travel. Both P-waves and S-waves travel through the Earth's interior. Surface waves, on the other hand, travel along the Earth's surface and are responsible for most of the damage caused by earthquakes. These waves are slower than P and S waves but have larger amplitudes and longer durations, causing more intense ground motion. The distinct characteristics of these wave types allow seismologists to analyze and understand the nature of earthquakes.

Measuring the magnitude and intensity of earthquakes provides critical information for understanding their power and impact. Magnitude, a measure of the energy released at the earthquake's source, is commonly determined using the moment magnitude scale (Mw). This scale, a successor to the Richter scale, is logarithmic, meaning each whole number increase represents a tenfold increase in amplitude and approximately 32 times more energy released. For example, a magnitude 7.0 earthquake releases about 32 times more energy than a magnitude 6.0 earthquake. Earthquake intensity, in contrast, measures the observed effects of an earthquake at a particular location. Intensity is influenced by factors such as distance from the epicenter, local geology, and building construction. The Modified Mercalli Intensity (MMI) scale, a twelve-level scale using Roman numerals, is often used to describe the severity of shaking at different locations. Understanding both magnitude and intensity provides a more comprehensive picture of an earthquake's effects.

Locating an earthquake requires analyzing data from multiple seismograph stations. Seismographs are sensitive instruments that detect and record ground motion caused by seismic waves. By comparing the arrival times of P-waves and S-waves at different stations, scientists can triangulate the earthquake's epicenter, the point on the Earth's surface directly above the hypocenter. The difference in arrival times between these wave types, known as the S-P interval, is proportional to the distance from the station to the epicenter. Using data from at least three stations, seismologists can pinpoint the epicenter with considerable accuracy. This information is crucial for emergency response and for understanding the regional tectonic activity.

Despite significant advancements in seismology, accurately predicting earthquakes remains a formidable challenge. While scientists can identify areas at high risk of earthquakes based on historical data and tectonic activity, pinpointing the exact time, location, and magnitude of a future

earthquake is currently not possible. Research continues on various potential precursors, such as changes in ground deformation, radon gas emissions, and animal behavior, but none have proven reliable for predicting earthquakes. Therefore, focusing on preparedness and mitigation strategies is essential for minimizing the impact of earthquakes.
Earthquake preparedness is vital for individuals and communities in earthquake-prone regions. Developing a comprehensive earthquake plan is the first step. This plan should include identifying safe places within your home, securing heavy objects that could fall, and establishing communication protocols with family members. Creating an emergency kit containing essential supplies like water, food, first aid supplies, flashlights, and a radio is crucial for surviving the aftermath of an earthquake. Participating in earthquake drills and familiarizing yourself with community evacuation plans can also enhance preparedness. Educating yourself about earthquake safety measures and staying informed about potential hazards can significantly reduce risks and contribute to community resilience. While earthquake prediction remains elusive, proactive preparedness can significantly mitigate their devastating impacts.

5.1 Faults & Seismic Waves

The Earth's crust isn't a single, solid piece, but rather a mosaic of immense fragments called tectonic plates. These plates, constantly in motion, albeit incredibly slow, interact with each other in various ways, creating the geological phenomena we experience as earthquakes and volcanoes. Critical to understanding earthquakes is the concept of faults, which are fractures in the Earth's crust where rocks have moved past each other. These movements aren't smooth and continuous; instead, they involve periods of stress buildup followed by sudden releases of energy. It's this sudden release of accumulated strain that generates the seismic waves we associate with earthquakes.
Faults come in three primary types: normal, reverse (also called thrust),

and strike-slip. Normal faults occur when the crust is extended, causing one block of rock to slide downwards relative to the other. Visualize this as pulling a rug apart, causing one side to drop lower. Reverse faults, conversely, are created by compressional forces, pushing one block upwards and over the other. Imagine pushing a rug against a wall, causing it to buckle and overlap itself. Strike-slip faults, the third type, involve horizontal movement, where the blocks slide past each other laterally. Picture two bricks sliding past each other on a tabletop. The San Andreas Fault in California is a prime example of a strike-slip fault, demonstrating this horizontal shearing motion.

When the built-up stress along a fault exceeds the strength of the surrounding rocks, the fault ruptures. This rupture initiates a cascade of energy released in the form of seismic waves that radiate outwards from the point of origin, known as the hypocenter or focus. These waves travel through the Earth's interior and along its surface, causing the ground to shake. Understanding the different types of seismic waves is crucial for interpreting earthquake behavior and the resulting damage patterns.

There are four main types of seismic waves: P-waves, S-waves, Love waves, and Rayleigh waves. P-waves, or primary waves, are compressional waves, similar to sound waves. They travel the fastest and can move through solids, liquids, and gases. Imagine a slinky being pushed and pulled – that's the motion of a P-wave. S-waves, or secondary waves, are shear waves. They move slower than P-waves and can only travel through solids. Think of shaking a rope up and down – that's the transverse motion of an S-wave.

Love waves and Rayleigh waves are surface waves, meaning they travel along the Earth's surface rather than through its interior. Love waves are horizontally polarized shear waves, meaning they move the ground side-to-side, perpendicular to the direction of wave propagation. They are particularly damaging to structures due to their strong shearing motion. Rayleigh waves, the slowest of the seismic waves, have a rolling motion, causing both vertical and horizontal ground movement. Imagine an ocean wave rolling in – that's similar to the motion of a Rayleigh wave. These surface waves are responsible for most of the shaking felt during an earthquake and consequently, the majority of the damage.

The speed at which these waves travel depends on the properties of the rocks they pass through. Denser rocks transmit seismic waves faster than less dense rocks. This difference in wave speeds allows seismologists to determine the location and depth of an earthquake. By analyzing the arrival times of different seismic waves at various seismograph stations around the world, scientists can triangulate the earthquake's hypocenter. Furthermore, the amplitude, or height, of the seismic waves, provides information about the magnitude, or size, of the earthquake.

The intricate interplay between faults and seismic waves forms the basis of our understanding of earthquakes. By studying fault types, their movement patterns, and the characteristics of the seismic waves they generate, we can gain valuable insights into the complex processes that shape our planet. This knowledge is not only essential for scientific advancement but also crucial for developing effective earthquake mitigation strategies and minimizing the risks associated with these powerful natural phenomena.

The interaction between the different types of waves and the Earth's varied geological structures provides a wealth of information for seismologists, helping them unravel the complexities of our dynamic planet.

5.2 Measuring Earthquakes

Understanding the power of an earthquake involves more than just witnessing its destructive aftermath. It requires quantifying the energy released, the ground motion produced, and the impact on both natural landscapes and human-built structures. This quantification is achieved through various scales and instruments, each designed to capture a different facet of these powerful events. Historically, intensity scales provided a qualitative assessment based on observed effects. One such scale, the Modified Mercalli Intensity Scale, categorizes earthquake intensity from I (not felt) to XII (catastrophic destruction). While useful for understanding local impact, intensity scales are subjective and vary depending on factors like building construction and population density. This subjectivity led to the development of magnitude scales, which aim for an objective measurement of the earthquake's energy release.

The Richter scale, developed in 1935 by Charles Richter, was a pioneering effort in this direction. It uses a logarithmic scale to quantify the amplitude of seismic waves recorded on seismographs. A one-unit increase on the Richter scale represents a tenfold increase in wave amplitude and approximately 32 times more energy released. While widely recognized, the Richter scale has limitations, particularly for very large earthquakes. Its measurements become saturated, meaning it underestimates the true magnitude of extremely powerful events. This limitation prompted the development of moment magnitude (Mw), which is now the preferred scale among seismologists. Moment magnitude considers the total energy released by an earthquake, calculated from the rupture area of the fault, the amount of slip, and the rigidity of the rocks. It provides a more accurate measure for large earthquakes and aligns better with the physical processes involved.

Modern seismographs are highly sensitive instruments capable of detecting even the slightest ground vibrations. These instruments are essential for measuring earthquake magnitude, locating their epicenters, and understanding the propagation of seismic waves. They work on the principle of inertia. A suspended mass remains relatively stationary while the ground and the rest of the instrument move during an earthquake. This relative motion is recorded electronically, providing a detailed record of the ground's shaking over time. These recordings, called seismograms, are a treasure trove of information for seismologists. They reveal not only the magnitude and location of an earthquake but also the different types of seismic waves generated—P-waves, S-waves, and surface waves—each traveling at different speeds and exhibiting distinct characteristics.

Analyzing seismograms from multiple stations allows scientists to pinpoint the earthquake's hypocenter, the precise point of rupture beneath the Earth's surface, and its epicenter, the point directly above it on the surface. This triangulation process relies on the time difference between the arrival of P-waves and S-waves at different seismic stations. The further a station is from the epicenter, the greater the time lag between the arrival of these two wave types. The precise location information provided by these analyses is crucial for understanding the distribution of earthquakes, identifying active fault zones, and assessing seismic hazards. Beyond magnitude and location, earthquake measurement extends to understanding the ground motion itself. Strong motion accelerometers are deployed in earthquake-prone regions to record the acceleration of the ground during shaking. This data is critical for engineers and structural designers, as it informs building codes and helps ensure structures can withstand the forces generated by strong earthquakes.

Furthermore, ground motion data contributes to the development of shake maps, which visually depict the intensity of shaking experienced across a region following an earthquake. Shake maps combine instrumental data with observations of damage and felt reports to provide a comprehensive view of the earthquake's impact. They are invaluable tools for emergency response, guiding rescue efforts and prioritizing areas needing immediate assistance. They also play a critical role in post-disaster assessments, helping to understand the extent of damage and inform recovery and reconstruction efforts. The study of earthquakes is a multifaceted endeavor. From basic intensity observations to sophisticated instrumental measurements, the tools and techniques employed provide a nuanced understanding of these powerful natural phenomena. By accurately measuring earthquakes, we gain valuable insights into the Earth's dynamics and improve our ability to mitigate the risks they pose to communities worldwide. This knowledge is paramount for developing effective preparedness strategies, designing resilient infrastructure, and ultimately safeguarding lives and livelihoods in earthquake-prone regions. The ongoing advancements in earthquake monitoring and analysis contribute to a more comprehensive understanding of these complex events, leading to more effective mitigation strategies and improved community resilience.

5.3 Locating Earthquakes

Pinpointing the precise location of an earthquake's origin, its hypocenter, is a complex yet crucial process in seismology. It relies on a global network of seismic stations, each equipped with sensitive instruments called seismographs that detect and record ground motion. These recordings, known as seismograms, provide the raw data needed to triangulate the earthquake's location. The key to this triangulation lies in the different speeds at which seismic waves travel through the Earth. Specifically, we utilize the difference in arrival times between P-waves (primary waves) and S-waves (secondary waves). P-waves, being compressional waves, travel faster and arrive at seismic stations first. S-waves, being shear waves, travel slower and arrive later. This time lag, known as the S-P interval, is directly proportional to the distance between the earthquake's hypocenter and the seismic station. A larger S-P interval indicates a greater distance.

Imagine a single seismic station. While it can determine the distance to an earthquake using the S-P interval, it cannot pinpoint the direction. The earthquake could be anywhere along a circle centered on the station, with a radius equal to the calculated distance. This is where the network of seismic stations comes into play. By analyzing the S-P intervals recorded at multiple stations, we can narrow down the earthquake's location. Each station provides a circle of possible locations, and the intersection of these circles reveals the earthquake's epicenter – the point on the Earth's surface directly above the hypocenter. Ideally, data from at least three seismic stations are required for accurate triangulation.

The process isn't simply a matter of drawing circles on a map, however. The Earth is not a perfectly homogeneous sphere, and seismic waves don't travel at constant speeds. Their velocity varies depending on the density and composition of the rocks they pass through. Consequently, the calculation must take into account the Earth's three-dimensional structure and the varying velocities of seismic waves. Sophisticated computer algorithms are used to model these variations and generate a more accurate location for the hypocenter. These algorithms consider factors such as the Earth's curvature, the different layers within the Earth, and even the effects of topography. They use iterative processes to refine the estimated location, minimizing the difference between the observed arrival times and the predicted arrival times based on the model.

Beyond simply locating the epicenter, determining the depth of the earthquake, or the focal depth, is also critical. This requires further analysis of the seismic waves, particularly those that reflect or refract off different layers within the Earth. Deep earthquakes, those originating hundreds of kilometers below the surface, typically occur at subduction zones where one tectonic plate dives beneath another. Shallow earthquakes, on the other hand, are more common and generally associated with transform faults or the upper parts of subduction zones. The depth of an earthquake significantly influences its impact at the surface. Shallow earthquakes, even those of moderate magnitude, can cause significant damage, while deeper earthquakes of the same magnitude tend to have a wider but less intense effect.

Modern seismic networks are increasingly incorporating real-time data processing and automated location algorithms. This allows for rapid determination of earthquake parameters, which is essential for issuing timely warnings and mobilizing emergency response efforts. The speed and accuracy of earthquake location have dramatically improved over the years. Today, the location of even relatively small earthquakes can be determined within minutes, providing valuable information for scientists, emergency managers, and the public. These rapid assessments, alongside ongoing research into earthquake precursors, play a vital role in mitigating the impact of these powerful natural phenomena. The intricate dance of seismic waves, recorded by a global network of sensitive instruments, allows us to unravel the mysteries hidden beneath the Earth's surface and better understand the forces that shape our world.

5.4 Earthquake Prediction

Accurately predicting earthquakes remains one of the most challenging frontiers in seismology. While we can identify regions at high risk due to their proximity to active fault lines, pinpointing the precise time, location, and magnitude of a future earthquake remains elusive. This difficulty arises from the complex and chaotic nature of the Earth's crust. The forces at play, the friction between rock faces, and the build-up and release of stress are incredibly intricate processes that are difficult to model with complete accuracy. Furthermore, the Earth's crust is not uniform, exhibiting variations in rock properties and stress levels, adding another layer of complexity to earthquake prediction. Despite these challenges, researchers continue to pursue a variety of approaches in hopes of eventually achieving reliable predictions.

One promising avenue involves closely monitoring precursory phenomena, subtle changes that might precede an earthquake. These include changes in groundwater levels, the release of radon gas from the Earth, and even slight deformations in the Earth's surface. Ground deformation can be measured using instruments like tiltmeters and strainmeters, which detect even minute shifts in the ground's slope or stretching. Similarly, changes in the electrical conductivity of rocks and variations in the Earth's magnetic field have been observed prior to some earthquakes, suggesting their potential as predictive indicators. While intriguing, none of these precursory signals have proven consistently reliable. Some earthquakes exhibit these changes while others do not, and conversely, these changes have occurred without subsequent earthquakes. This inconsistency highlights the complex relationship between these phenomena and the actual rupture process.

Statistical analysis of past earthquake patterns forms another approach to prediction. By studying the historical record of earthquakes along a particular fault, seismologists can sometimes identify recurring patterns in the timing and size of earthquakes. This approach, known as probabilistic seismic hazard assessment, allows scientists to estimate the likelihood of an earthquake of a certain magnitude occurring within a specified timeframe and region. While this method does not provide specific predictions for individual earthquakes, it provides crucial information for long-term planning and risk mitigation. Building codes, insurance rates, and emergency preparedness plans can be tailored based on these probabilistic assessments. This approach recognizes the inherent uncertainty in earthquake prediction while offering valuable insights into long-term seismic risks.

Understanding the physics of earthquake rupture is also crucial for improving prediction capabilities. Laboratory experiments on rock samples subjected to high pressures and temperatures help scientists understand how rocks behave under stress and how fractures initiate and propagate. These experiments, combined with sophisticated computer models, are slowly uncovering the complex interactions that lead to earthquakes. These models simulate the behavior of faults under different stress conditions, allowing researchers to study the factors that influence rupture nucleation and propagation. Furthermore, advancements in high-performance computing allow for increasingly detailed simulations that incorporate more realistic geological complexities.

Advances in monitoring technologies, particularly through dense networks of seismometers, provide increasingly detailed data on ground motion and seismic activity. These dense networks can detect even the smallest tremors, known as micro-earthquakes, which might provide insights into the stress buildup on a fault. Analyzing the patterns and distribution of these micro-earthquakes can help identify areas where stress is accumulating and potentially pinpoint zones with a higher probability of future rupture. This detailed monitoring also helps refine our understanding of fault geometry and the distribution of stress along the fault zone. This refined understanding is crucial for improving the accuracy of earthquake

prediction models and simulations.
Integrating various approaches, combining precursory signal monitoring, statistical analysis, physical modeling, and advanced monitoring technologies, offers the most promising path forward. A multi-faceted approach acknowledges the complex and multifaceted nature of earthquake processes. By combining data from diverse sources, researchers can build a more comprehensive understanding of the factors leading to earthquakes. While truly accurate earthquake prediction remains a long-term goal, the ongoing research and technological advancements provide hope for eventually achieving reliable predictions, ultimately saving lives and mitigating the devastating impact of these powerful natural events. The pursuit of accurate earthquake prediction is not merely a scientific endeavor but a critical step towards enhancing community resilience and preparedness in earthquake-prone regions.

5.5 Earthquake Preparedness

Securing your safety during an earthquake involves a multi-pronged approach, encompassing preparation before, action during, and recovery after the event. A well-defined plan can significantly reduce risks to yourself and your loved ones. Start by assessing your home for potential hazards. Secure heavy furniture to walls using brackets or straps. This prevents bookcases, dressers, and appliances from toppling over during tremors. Mirrors and picture frames hung above beds or seating areas should be relocated or secured with specialized hardware. Latches on cupboard doors will prevent contents from spilling out, minimizing potential injuries and cleanup. Identify safe spaces within your home. Sturdy tables, desks, and interior doorways offer some protection from falling debris. Stay away from windows and exterior walls during shaking.
Next, assemble an emergency kit. This kit should contain enough supplies to sustain you and your family for at least 72 hours, possibly longer. Include essential items like bottled water, non-perishable food, a first-aid kit,

a battery-powered or hand-crank radio, flashlights with extra batteries, a whistle to signal for help, and a multi-tool or Swiss Army knife. Personal hygiene items, medications, and important documents (copies of identification, insurance policies, and medical records) should also be included. Keep this kit in an easily accessible location, known to all members of your household. Consider a smaller, portable kit for your car and workplace.

Developing a communication plan is crucial. Choose an out-of-state contact person who can serve as a central communication hub for family members. Earthquake disruptions can make local communication difficult, so having a single point of contact outside the affected area can help ensure everyone can connect. Teach family members how to use text messaging or social media to communicate, as these methods often function even when phone lines are down. Practice your earthquake drill regularly. This familiarizes everyone with the procedures, increasing the likelihood of a calm and effective response during a real event.

When an earthquake strikes, the immediate priority is to "Drop, Cover, and Hold On." Drop to the ground on your hands and knees. This position protects you from being knocked over and allows you to crawl to a safe space if needed. Cover your head and neck with your arms and seek shelter under a sturdy table or desk. If no shelter is available, crawl next to an interior wall and cover your head and neck. Hold on to your shelter or your head and neck until the shaking stops. Be prepared for aftershocks, which can occur minutes, hours, or even days after the initial earthquake.

Once the shaking subsides, carefully assess your surroundings. Check for injuries and provide first aid if necessary. Be aware of potential hazards like broken glass, downed power lines, and gas leaks. If you smell gas or hear a hissing sound, leave the area immediately and report it to authorities. Use your battery-powered radio or other communication devices to stay informed about emergency instructions and updates. Avoid using your phone for non-essential calls to keep lines clear for emergency responders. If your home is damaged, do not re-enter until it has been inspected by authorities and deemed safe.

The recovery phase after an earthquake can be lengthy and challenging. Be patient and cooperative with emergency personnel. Document any damage to your property for insurance purposes. Reach out to your insurance company and follow their procedures for filing a claim. If your home is uninhabitable, seek temporary shelter with friends, family, or at a designated evacuation center. The emotional toll of an earthquake can be significant. Seek support from family, friends, or mental health professionals if needed. Remember, preparedness is the key to navigating these natural events safely. By taking proactive steps and developing a comprehensive plan, you can significantly reduce the risks and impact of earthquakes on your life and community.

6 Chapter 6: Tectonic Plate Boundaries

Our planet's surface is a dynamic mosaic of constantly shifting plates, a concept central to understanding the distribution of volcanoes and earthquakes. These plates, fragments of the lithosphere (the Earth's crust and uppermost mantle), interact at distinct boundaries, each characterized by unique geological processes. These interactions are the driving force behind the majority of the Earth's seismic and volcanic activity, shaping our world in profound ways. We categorize these interactions into three primary types: divergent, convergent, and transform boundaries.

Divergent boundaries represent areas where tectonic plates are moving apart from each other. This separation creates a rift, or gap, in the Earth's crust, allowing molten rock from the mantle, known as magma, to rise and fill the void. As the magma cools and solidifies, new crust is formed, effectively pushing the plates further apart. This process, known as seafloor spreading, is most evident along mid-ocean ridges, vast underwater mountain ranges that snake across the ocean floor. Iceland, situated directly atop the Mid-Atlantic Ridge, offers a unique opportunity to witness this process on land, marked by frequent volcanic eruptions and geothermal activity. The continuous creation of new crust at divergent boundaries explains the relatively young age of the oceanic crust compared to the much older continental crust. The slow but relentless spreading also contributes to the gradual shifting of continents over geological timescales.

Convergent boundaries, in contrast to divergent boundaries, are zones where tectonic plates collide. These collisions are powerful geological events with far-reaching consequences. When an oceanic plate meets a continental plate, the denser oceanic plate subducts, or dives beneath, the lighter continental plate. This subduction process creates a deep oceanic trench and triggers volcanic activity along the edge of the overriding continental plate. The Cascade Range in the Pacific Northwest of the United States is a prime example of this type of convergent boundary, characterized by a chain of volcanoes fueled by the subduction of the Juan de Fuca plate beneath the North American plate. When two oceanic plates collide, one typically subducts beneath the other, forming a volcanic island arc, a chain of volcanic islands that parallels the trench. Japan and the Philippines are classic examples of island arcs formed by this process. Finally, when two continental plates collide, the immense pressure and friction cause the crust to buckle and thicken, creating towering mountain ranges like the Himalayas. These continental collisions are associated with powerful earthquakes but typically lack volcanic activity due to the absence of subduction.

Transform boundaries represent the third type of plate interaction, where plates slide horizontally past each other. Unlike divergent and convergent boundaries, transform boundaries neither create nor destroy crust. They are characterized by intense shearing forces that generate significant seismic activity. The San Andreas Fault in California is perhaps the most well-known example of a transform boundary, where the Pacific plate slides past the North American plate. The movement along transform boundaries is not smooth and continuous but rather occurs in a series of sudden slips, releasing accumulated stress in the form of earthquakes. These earthquakes can range from minor tremors to devastating events, depending on the magnitude of the slip and the depth at which it occurs. While volcanoes are generally absent along transform boundaries, the intense fracturing of the crust can create pathways for magma to reach the surface in some instances, leading to localized volcanic activity.

Understanding the interplay of these three boundary types – divergent, convergent, and transform – is essential for comprehending the complex distribution of volcanoes and earthquakes across the globe. These boundaries are not static features but rather dynamic zones of constant interaction, continuously reshaping the Earth's surface and driving the geological processes that shape our planet. The study of plate tectonics provides a framework for interpreting the past, understanding the present, and, to some extent, predicting the future of Earth's geological activity. The locations of volcanoes and earthquakes are not random; they are intimately linked to the movements and interactions of these tectonic plates. By studying the characteristics of each boundary type and the geological features associated with them, we can gain a deeper appreciation for the powerful forces at work beneath our feet. This knowledge is crucial not only for scientific understanding but also for mitigating the risks associated with these natural hazards.

6.1 Divergent Boundaries

Imagine two giant conveyor belts, facing each other but moving apart. That's the basic idea behind divergent plate boundaries. These are zones where the Earth's lithospheric plates are pulling away from one another. This pulling action, driven by convection currents in the mantle below, creates a rift, a linear zone where the lithosphere is thinning. Magma, molten rock from the mantle, wells up to fill the void. As the plates continue to separate, this magma erupts onto the surface, solidifying and forming new oceanic crust. This continuous process of seafloor spreading steadily widens the gap between the diverging plates, gradually expanding ocean basins over millions of years. Think of it as the Earth's way of renewing its skin, constantly creating fresh oceanic crust while older crust is pushed away from the rift.

A classic example of a divergent boundary is the Mid-Atlantic Ridge, an underwater mountain range that snakes its way down the center of the

Atlantic Ocean. Along this ridge, new oceanic crust is constantly being generated, pushing North America and Europe further apart. The ridge itself is a testament to the immense power of these geological processes, marked by volcanic activity, hydrothermal vents teeming with life, and frequent, albeit mostly minor, earthquakes. These earthquakes, while usually not as devastating as those at convergent boundaries, provide valuable insights into the ongoing processes of plate separation and magma upwelling. They are a constant reminder of the dynamic forces at play beneath the ocean's surface.

Divergent boundaries don't just exist under the sea. They also occur on land, forming what are known as continental rift valleys. The East African Rift Valley is a prime example, a vast geological depression stretching thousands of kilometers. Here, the African plate is slowly tearing apart, creating a landscape characterized by volcanoes like Mount Kilimanjaro and Mount Kenya, hot springs, and deep lakes. Over millions of years, this rifting process may eventually split the African continent, forming a new ocean basin. Observing this active rift provides us with a unique opportunity to witness the early stages of continental breakup, a process that has shaped our planet's surface throughout its history.

The process of divergence begins with upwelling magma from the mantle. This magma doesn't simply surge to the surface. It creates a series of interconnected fractures and fissures in the overlying crust. These cracks act as pathways for magma to intrude into the brittle lithosphere. The intense heat from the magma weakens the surrounding rocks, further facilitating the rifting process. As the plates pull apart, the crust thins and subsides, forming a rift valley. The magma continues to rise, erupting onto the surface as lava flows and building volcanic landforms along the rift margins. This continuous interplay of magmatism, faulting, and crustal thinning is the essence of a divergent boundary.

The type of volcanism at divergent boundaries differs significantly from that seen at convergent boundaries. At divergent boundaries, the magma is typically basaltic, meaning it is low in silica and has a low viscosity. This allows the magma to flow easily, resulting in effusive eruptions characterized by gentle lava flows rather than explosive blasts. The volcanic rocks formed at divergent boundaries are typically dark-colored and dense, reflecting the basaltic composition of the magma. The lava flows can spread over vast areas, creating extensive lava plateaus and gently sloping shield volcanoes. These broad, dome-shaped volcanoes, like Mauna Loa in Hawaii, are built up layer by layer by successive lava flows.

The rate of spreading at divergent boundaries varies significantly. The Mid-Atlantic Ridge, for instance, spreads at a rate of about 2.5 centimeters per year, roughly the same speed as your fingernails grow. The East Pacific Rise, on the other hand, spreads much faster, at rates exceeding 15 centimeters per year. This difference in spreading rates affects the topography of the ridge. Slow-spreading ridges tend to have steep, rugged terrain, while fast-spreading ridges are characterized by broader, smoother topography. The spreading rate also influences the frequency and intensity of volcanic activity and earthquakes along the boundary.

Understanding divergent boundaries is crucial for comprehending the larger picture of plate tectonics. They represent a fundamental process that shapes the Earth's surface, creates new oceanic crust, and drives the movement of continents. Studying these boundaries, whether beneath the ocean waves or in the heart of a continental rift, provides us with valuable insights into the dynamic forces that have shaped our planet for billions of years and continue to sculpt its future. By examining the rocks formed at divergent boundaries, analyzing the patterns of earthquakes, and monitoring the movement of the plates, scientists can piece together the complex story of Earth's ever-evolving surface. This knowledge is not just academic; it helps us understand the distribution of natural resources, assess the risks of geological hazards, and appreciate the intricate workings of our dynamic planet.

6.2 Convergent Boundaries

Imagine two colossal tectonic plates, driven by the immense forces within the Earth, inexorably moving towards each other. This collision zone, a convergent boundary, is a place of intense geological activity, responsible for creating some of the planet's most dramatic landscapes and catastrophic events. Here, the Earth's crust buckles, folds, and fractures, giving rise to towering mountain ranges, deep ocean trenches, powerful earthquakes, and explosive volcanoes. Understanding the dynamics of convergent boundaries is crucial to grasping the forces that shape our world.

Convergent boundaries are classified into three main types based on the nature of the colliding plates: oceanic-oceanic, oceanic-continental, and continental-continental. Each type of collision produces unique geological features and hazards. Consider, first, the meeting of two oceanic plates. The denser of the two plates will subduct, or dive beneath, the other, plunging into the Earth's mantle. This subduction process creates a deep oceanic trench, a dramatic scar on the ocean floor marking the point of descent. As the descending plate sinks and heats up, it releases water into the surrounding mantle. This water lowers the melting point of the mantle rock, generating magma. This buoyant magma rises to the surface, erupting to form volcanic island arcs, chains of volcanic islands that parallel the trench. Examples of these island arcs include Japan, the Philippines, and the Aleutian Islands. The friction between the subducting plate and the overriding plate also generates powerful earthquakes along the subduction zone, sometimes triggering devastating tsunamis.

When an oceanic plate collides with a continental plate, a similar process of subduction occurs. The denser oceanic plate dives beneath the lighter continental plate, creating a deep oceanic trench offshore. Again, the subducting plate releases water, triggering magma formation in the mantle. This magma rises and erupts, creating a chain of volcanoes along the edge of the continent, known as a continental volcanic arc. The Andes Mountains in South America, with their towering volcanoes, are a prime example of this type of convergent boundary. The Cascade Range in the Pacific Northwest of the United States also demonstrates this phenomenon. The subduction zone between the Juan de Fuca oceanic plate and the North American continental plate creates the volcanic peaks of Mount Rainier, Mount St. Helens, and Mount Hood, along with frequent earthquakes.

The collision of two continental plates creates a very different scenario. Continental crust is too buoyant to subduct easily. Instead, the colliding plates crumple and thicken, uplifting the crust to form towering mountain ranges. The Himalayas, the highest mountain range on Earth, is a dramatic testament to this continental collision. The Indian plate continues to push northward into the Eurasian plate, causing the Himalayas to rise higher each year. This ongoing collision generates powerful earthquakes, a constant reminder of the immense forces at work. The Alps in Europe and the Appalachians in the eastern United States are also products of ancient continental collisions. While volcanoes are less common at continental-continental convergent boundaries, the heat and pressure generated by the collision can cause metamorphism, transforming existing rocks into new types.

Understanding the interplay of these forces is essential for appreciating the dynamism of our planet. Convergent boundaries are not static features. They are constantly evolving, driven by the slow but relentless movement of the tectonic plates. These dynamic zones are responsible for some of Earth's most spectacular landscapes and most devastating natural disasters. The towering peaks of the Himalayas, the fiery eruptions of the Andes volcanoes, and the devastating earthquakes along the Pacific Ring of Fire are all manifestations of the powerful forces unleashed at convergent boundaries. By studying these boundaries, we gain invaluable insights into the ongoing processes that shape our world, contributing to a deeper understanding of Earth's complex and ever-changing nature. The continuing study of these regions allows scientists to monitor changes, improve predictive models, and implement mitigation strategies to protect vulnerable populations from the hazards associated with these powerful geological processes. This ongoing research deepens our understanding of Earth's dynamics and helps us to live more safely on our active planet.

6.3 Transform Boundaries

Imagine two colossal tectonic plates, not colliding head-on or drifting apart, but grinding relentlessly against each other. This lateral movement, where plates slide horizontally, defines transform boundaries. These boundaries are not marked by the fiery spectacle of volcanic eruptions or the creation of towering mountain ranges, but they are a significant source of seismic activity, frequently generating powerful earthquakes. The friction between the plates builds up immense pressure, and when this pressure is finally released, it unleashes tremendous energy in the form of seismic waves, shaking the Earth's crust.

The most famous example of a transform boundary is the San Andreas Fault in California. This fault marks the boundary between the Pacific Plate and the North American Plate. The Pacific Plate, carrying the vast expanse of the Pacific Ocean floor, inches northwestward relative to the North American Plate at a rate of approximately 5 centimeters per year, a seemingly insignificant pace but one that has profound geological consequences. This creeping movement doesn't occur smoothly; instead, the plates lock together due to friction, storing elastic potential energy like a tightly wound spring. Over decades, even centuries, the strain accumulates until the locked sections of the fault rupture, releasing the stored energy in a sudden, violent jolt—an earthquake.

While the San Andreas Fault is perhaps the most well-known, transform boundaries are not limited to continental crust. They also occur on the ocean floor, creating a network of fracture zones that dissect the mid-ocean ridges. These underwater transform faults play a critical role in the overall process of plate tectonics. They accommodate the different spreading rates along segments of the mid-ocean ridges, allowing the plates to move apart smoothly despite variations in the speed of seafloor spreading. These underwater faults, though less visible than their terrestrial counterparts, contribute significantly to the global seismic activity map. They produce earthquakes that, while often occurring deep beneath the ocean's surface, can still generate significant tremors and, in certain circumstances, even tsunamis if the seafloor is displaced vertically.

The movement along transform boundaries is not always a steady creep. Sometimes, it occurs in a series of sudden jerks, known as stick-slip behavior. Imagine pushing a heavy box across a rough floor. It might stick for a while, building up friction, and then suddenly slip forward. This stick-slip motion is analogous to the behavior of rocks along a transform fault. The "stick" phase corresponds to the period of stress buildup, where the plates are locked together. The "slip" phase represents the earthquake itself, the sudden release of accumulated energy. The size of the earthquake depends on the area of the fault that ruptures and the amount of slip that occurs. A small rupture with minimal slip will produce a minor tremor, while a large rupture with significant slip can generate a devastating earthquake.

The landscape along transform boundaries often reflects the powerful forces at play. Long, linear valleys, offset streams, and scarps – steep, cliff-like faces – are common features. These landforms are the direct result of the horizontal shearing motion. For instance, a river flowing across a transform fault might be offset over time, creating a distinct bend in its course. Similarly, repeated earthquakes can create scarps where the ground has been vertically displaced. These features, although often subtle, provide visual evidence of the ongoing tectonic activity. Geologists use these surface expressions to map the fault's trace and to study its past behavior, gaining insights into the frequency and magnitude of earthquakes that have occurred along the fault over geological time. This information is crucial for assessing earthquake hazards and developing mitigation strategies.

While transform boundaries are primarily associated with earthquakes, they can also exhibit secondary geological features. In some cases, localized compression or tension can occur along bends or irregularities in the fault, leading to the formation of small hills or basins. Additionally, the intense grinding and fracturing of rocks along the fault zone can create highly fractured and permeable rock, creating pathways for groundwater flow and sometimes resulting in springs or seeps along the fault trace. Understanding the intricacies of these secondary features provides further insights into the complex dynamics of transform boundaries and their impact on the surrounding environment. Their study deepens our understanding of the Earth's dynamic crust and enhances our ability to predict and prepare for future seismic events.

7 Chapter 7: Earthquake Impacts

The earth trembles. Buildings sway, then crumble. Roads crack and split, bridges twist and fall. This is the immediate, visceral power of ground shaking, the most direct impact of an earthquake. The intensity of this shaking is dictated by a complex interplay of factors: the magnitude of the earthquake, the distance from the epicenter, the local geology, and even the depth of the quake's focus. Bedrock, for instance, transmits seismic waves more efficiently than softer sediments, leading to less intense shaking in areas built upon solid rock. Conversely, areas situated on loose, unconsolidated soils, particularly those saturated with water, are susceptible to a phenomenon called liquefaction. In such instances, the ground loses its strength and behaves like a liquid, causing buildings to sink and underground infrastructure to fail catastrophically. Understanding these geological nuances is critical in urban planning and building design, influencing everything from building codes to the choice of construction materials. Seismic engineering strives to create structures that can withstand the stresses of ground shaking, employing techniques like base isolation and damping systems to minimize damage.

Beyond the immediate devastation of collapsing structures, ground shaking triggers a cascade of secondary impacts. Landslides, often devastating in mountainous regions, can be unleashed by even moderate earthquakes. The shaking destabilizes slopes, sending tons of rock and debris hurtling down mountainsides, burying homes and blocking vital transportation routes. Similarly, ground shaking can disrupt the delicate balance of underground aquifers, causing changes in groundwater levels. Wells might suddenly dry up, or conversely, new springs might emerge. The disruption to water resources can have profound impacts on agriculture and access to clean drinking water, exacerbating the humanitarian crisis that often follows a major earthquake. Furthermore, the rupture of gas lines and electrical cables during ground shaking ignites fires, adding another layer of complexity to rescue and recovery efforts. The 1906 San Francisco earthquake provides a stark example of this, where fires caused more damage than the initial shaking. These secondary effects highlight the interconnectedness of earthquake impacts and the need for a holistic approach to disaster preparedness.

While ground shaking represents the most immediate and localized threat from earthquakes, tsunamis possess the terrifying potential to extend devastation far beyond the earthquake's origin. Generated by the rapid displacement of a massive volume of water, typically caused by undersea earthquakes at subduction zones, tsunamis radiate outwards in all directions, traveling across vast ocean basins at incredible speeds. In the deep ocean, a tsunami might be barely perceptible, a gentle swell barely a meter high moving at hundreds of kilometers per hour. However, as these waves approach shallower coastal waters, they slow down, their energy compressing into a towering wall of water that can reach heights of tens of meters. The impact of a tsunami is catastrophic. Coastal communities are inundated, buildings are swept away, and the force of the water carries debris inland, transforming the landscape and leaving behind a trail of destruction.

The destructive power of a tsunami stems not only from its immense height but also from its incredible momentum. Unlike wind-driven waves that primarily affect the surface, tsunami waves involve the entire water column, carrying a tremendous amount of energy. This translates into a powerful surge that can penetrate far inland, overwhelming coastal defenses and carrying debris, vehicles, and even buildings far from the shoreline. The 2004 Indian Ocean tsunami, triggered by a magnitude 9.1 earthquake off the coast of Sumatra, provided a tragic demonstration of the devastating reach of these waves. The tsunami impacted coastlines across the Indian Ocean, from Indonesia to East Africa, claiming hundreds of thousands of lives and causing widespread destruction. This event underscored the importance of early warning systems and international cooperation in mitigating the impact of tsunamis.

Beyond the immediate loss of life and property, tsunamis leave a lasting legacy of environmental damage. Saltwater intrusion contaminates freshwater sources, rendering them unusable for drinking and agriculture. The powerful surge uproots trees and vegetation, stripping coastal ecosystems bare and increasing the risk of erosion. Debris carried by the tsunami pollutes beaches and coastal waters, impacting marine life and disrupting delicate ecological balances. Furthermore, the destruction of infrastructure, including sanitation systems, can lead to the spread of disease in the aftermath of a tsunami. The long-term recovery from a tsunami involves not only rebuilding homes and infrastructure but also restoring the natural environment and addressing the long-term health and social consequences. The complex interplay of physical, ecological, and societal impacts emphasizes the need for a comprehensive approach to tsunami preparedness and mitigation, incorporating scientific understanding, engineering solutions, and community engagement.

7.1 Ground Shaking

The earth shudders. Buildings sway. Bridges twist. This is the raw power of ground shaking, the most direct and devastating effect of an earthquake. Understanding the complexities of ground shaking is critical for mitigating its destructive potential. The intensity of ground shaking isn't uniform. It's influenced by a complex interplay of factors, starting with the magnitude of the earthquake itself. A higher magnitude earthquake releases more energy, leading to more violent shaking over a wider area. However, magnitude alone doesn't tell the whole story.

The distance from the epicenter, the point on the Earth's surface directly above the earthquake's origin, plays a crucial role. Shaking generally decreases with distance, but local geological conditions can significantly modify this pattern. Soft sediments, like those found in valleys and basins, amplify seismic waves, leading to stronger shaking than on solid bedrock. This phenomenon, known as site amplification, can dramatically increase the destructiveness of an earthquake in certain areas. Imagine shaking a bowl of jelly compared to a block of wood – the jelly will wobble much more intensely.

The type of fault movement also influences the characteristics of ground shaking. Different types of faults, such as strike-slip, normal, and reverse faults, produce different patterns of ground motion. For example, strike-slip faults, where two blocks of crust slide horizontally past each other, often generate strong lateral shaking, which can be particularly damaging to structures not designed to withstand sideways forces. Understanding these different fault mechanisms allows engineers to design buildings that can better resist the specific types of shaking expected in a given region.

Furthermore, the duration of shaking is a critical factor in determining the extent of damage. A longer period of shaking allows more energy to be transferred to structures, increasing the likelihood of collapse. This is why even moderate earthquakes that produce prolonged shaking can cause significant damage, while short, sharp jolts, even if intense, might have less impact. The duration is influenced by the size of the fault rupture and the complexity of the earthquake itself. Larger ruptures typically result in longer shaking durations.

Ground shaking manifests in different ways, each with its own destructive potential. Vertical shaking, a direct up-and-down motion, can cause buildings to bounce, potentially leading to structural failure if the building's natural frequency matches the frequency of the seismic waves. This phenomenon, known as resonance, can amplify the shaking, causing catastrophic damage. Think of pushing a child on a swing – timed correctly, even small pushes can create a large swing arc.

Horizontal shaking, the back-and-forth motion, is often the most damaging type of ground movement. It can cause buildings to sway and collapse, bridges to twist and buckle, and pipelines to rupture. This type of shaking is particularly dangerous for unreinforced masonry structures, which are common in many parts of the world. The lateral forces exerted by horizontal shaking can easily overcome the weak bonding between bricks or stones, leading to widespread destruction.

Rotational ground motion, though less common, can be extremely destructive. This twisting motion can tear apart foundations, causing buildings to tilt and collapse. It's particularly damaging to tall structures and those with asymmetrical designs. Imagine twisting the base of a tall tower – the upper floors would experience amplified movement, increasing the stress on the structure.

Finally, surface waves, which travel along the Earth's surface, can cause significant ground shaking, especially at greater distances from the epicenter. These waves, including Rayleigh waves (rolling motion) and Love waves (side-to-side motion), are slower than body waves (P and S waves) but carry a significant amount of energy. They can cause extensive damage to structures and infrastructure, particularly in areas with soft soil conditions. These waves are like ripples in a pond, spreading outwards and causing disturbances even far from the initial impact. Understanding the complex nature of ground shaking – its dependence on magnitude, distance, geology, fault type, duration, and the various types of ground motion – is crucial for developing effective earthquake-resistant building practices and mitigating the devastating impact of these powerful natural events.

7.2 Tsunamis

The immense power of earthquakes can unleash a far-reaching and devastating secondary hazard: tsunamis. These are not tidal waves, as they are sometimes misnamed, but rather a series of ocean waves generated by sudden, large-scale displacements of water. Most commonly, this displacement is caused by underwater earthquakes, particularly those occurring at subduction zones, where one tectonic plate dives beneath another. Landslides, volcanic eruptions, and even meteorite impacts can also generate tsunamis, though these are less frequent. The key factor is the rapid vertical movement of a large volume of water.

Understanding the mechanics of tsunami generation is crucial. Imagine an underwater earthquake rupturing a fault along a subduction zone. The sudden uplift of the seafloor displaces the water column above it, creating a bulge on the ocean surface. Gravity then acts on this bulge, pulling it down and outwards, initiating a series of waves that propagate across the ocean. These waves have extremely long wavelengths, sometimes hundreds of kilometers, and can travel at speeds comparable to a jetliner, over 800 kilometers per hour in the deep ocean. In deep water, the wave height is relatively small, often less than a meter, making them barely noticeable to ships at sea. However, as these waves approach shallower coastal waters, their speed decreases, causing the wavelength to compress and the wave height to increase dramatically.

This shoaling effect, combined with the focusing of wave energy as it enters bays and harbors, can result in towering waves that inundate coastal areas. The first wave of a tsunami is not always the largest. Sometimes the trough, or the low point between waves, arrives first, causing a dramatic drawback of the shoreline, exposing areas normally underwater. This drawback is a crucial warning sign of an impending tsunami and should be treated as an immediate signal to evacuate to higher ground. The actual tsunami waves can then arrive minutes or even hours later, crashing onto the shore with tremendous force.

The destructive power of a tsunami lies not just in its height but also in its incredible momentum. The immense volume of water carried by these waves carries debris, including uprooted trees, vehicles, and even buildings, transforming them into battering rams. The backwash, the receding water carrying debris back out to sea, can be equally devastating, dragging people and objects out into the ocean. The repeated surge and retreat of multiple waves can cause widespread destruction, eroding coastlines, destroying infrastructure, and claiming lives.

Mitigation strategies for tsunamis focus on early warning systems, coastal defenses, and community education. Deep-ocean Assessment and Reporting of Tsunamis (DART) buoys are strategically placed in the Pacific and Indian Oceans to detect tsunami waves and relay information to warning centers. These buoys measure pressure changes on the seafloor, providing valuable data for forecasting tsunami arrival times and potential impacts. Coastal communities utilize seawalls, breakwaters, and vegetation barriers to dissipate wave energy and reduce inundation. However, these structures are not always effective against the largest tsunamis.

Public awareness and education are paramount. Understanding the warning signs, such as earthquake shaking or a rapid ocean recession, and knowing evacuation routes and procedures can save lives. Regular tsunami drills and community preparedness programs are essential for building resilience against these devastating events. Tsunami warnings are disseminated through various channels, including sirens, radio broadcasts, television alerts, and mobile phone notifications. Heeding these warnings and taking immediate action is crucial for minimizing casualties and damage.

Beyond immediate evacuation, long-term planning and land use management play a vital role in reducing tsunami risk. Restricting development in high-risk coastal areas, designing buildings to withstand tsunami forces, and preserving natural coastal buffers like mangrove forests can mitigate the impacts of future tsunamis. The science of tsunami modeling and forecasting continues to advance, allowing for more accurate predictions and targeted warnings. By combining scientific understanding with effective preparedness measures, communities can strive to minimize the devastating impacts of these powerful ocean waves. Recognizing the potential for secondary hazards like tsunamis is a critical component of earthquake preparedness and disaster risk reduction strategies worldwide. Understanding the mechanisms of tsunami generation, propagation, and impact empowers communities to make informed decisions and build more resilient coastal regions in the face of this formidable natural hazard.

www.ingramcontent.com/pod-product-compliance
Lightning Source LLC
LaVergne TN
LVHW022024200926
848481LV00007B/2537
* 9 7 9 8 3 4 8 4 8 9 3 3 5 *